DIGITAL LITERACIES

Colin Lankshear, Michele Knobel,
and Michael Peters
General Editors

Vol. 30

PETER LANG
New York • Washington, D.C./Baltimore • Bern
Frankfurt am Main • Berlin • Brussels • Vienna • Oxford

DIGITAL LITERACIES

Concepts, Policies and Practices

Colin Lankshear & Michele Knobel, Editors

PETER LANG
New York • Washington, D.C./Baltimore • Bern
Frankfurt am Main • Berlin • Brussels • Vienna • Oxford

Library of Congress Cataloging-in-Publication Data

Digital literacies: concepts, policies and practices /
edited by Colin Lankshear, Michele Knobel.
p. cm. — (New literacies and digital epistemologies ; v. 30)
Includes bibliographical references and indexes.
1. Internet literacy. 2. Computer literacy. 3. Information literacy.
I. Knobel, Michele.
TK5105.875.I57D546   004—dc22   2008009678
ISBN: 978-1-4331-0168-7 (hardcover)
ISBN 978-1-4331-0169-4 (paperback)
ISSN 1523-9543

Bibliographic information published by **Die Deutsche Bibliothek**.
Die Deutsche Bibliothek lists this publication in the "Deutsche
Nationalbibliografie"; detailed bibliographic data is available
on the Internet at http://dnb.ddb.de/.

Cover design by Clear Point Designs

The paper in this book meets the guidelines for permanence and durability
of the Committee on Production Guidelines for Book Longevity
of the Council of Library Resources.

Printed in the United States of America

Contents

Acknowledgments

The editors wish to that the following colleagues, institutions, and friends for contributing in varying ways to making this book possible.

We particularly want to thank the chapter authors for their work, which has been undertaken in addition to their many other commitments. Their willingness to contribute has made it possible to produce a volume that ranges over a wide range of perspectives and approaches—which is precisely what a book on digital literacies needs to do at this time. We know how much time and effort are involved in producing chapters for collections and greatly appreciate the authors' generous participation.

The support of the various institutions employing the editors and authors is also appreciated. We want especially to thank Montclair State University, James Cook University, McGill University, Mount St Vincent University, and Central Queensland University for supporting our recent and ongoing projects.

We also thank all the folk at Peter Lang for supporting the New Literacies series generally and for their contributions to this volume in particular. Special thanks go to Bernadette Shade, Valerie Best, and Chris Myers, who have contributed with their trademark enthusiasm and good humor.

We want to pay special thanks to Chris Barrus for allowing us to use his Facebook profile as an in-depth case of participation in online social networking spaces, and for reading and commenting on drafts of our account of his profile. We are also grateful to John Lambie for permission to use his Facebook account of "how-we-met", to liebemarlene for permissions to reproduce material from her weblog and photostreams in Chapter 10, and to the many anonymous research participants in several countries who have informed studies underlying the work in this book.

Melinda Starc has contributed generously to formatting, proof reading, and general administrative work throughout the production of this book, and we gratefully acknowledge her contributions, which are always "on time" and "right first time".

We want to thank Lawrence Lessig for giving us permission to remix elements of his written work and oral presentations on remix and free culture. We also greatly appreciate the ways he has informed our thinking and writing about digital literacies, and esteem his unstinting activity on behalf of the free and responsible development of internet culture.

Finally, we want to acknowledge our appreciation of permissions to reproduce work, as follows:

The Nordic Journal of Digital Literacy, for David Buckingham (2006), Defining digital literacy: What do young people need to know about digital media? *The Nordic Journal of Digital Literacy* Vol. 1, No. 4: 263–276.

E-Learning, for Genevieve Johnson (2007), Functional internet literacy: Required cognitive skills with implications for instruction, *E-Learning* Vol. 4, No. 4: 433–441.

Nicholas Diakopoulos, for Figure 8.1, "Graphic representation of different modes of remix a they relate to people and media elements", reproduced from his (2005) Remix culture: Mixing up authorship, available at http://www.deakondesign.com/

Introduction

Digital Literacies—Concepts, Policies and Practices

COLIN LANKSHEAR AND MICHELE KNOBEL

This book supports an emerging trend toward emphasizing the plurality of digital literacy; recognizing the advantages of understanding digital literacy as digital *literacies*. In the book world this trend is still marginal. In December 2007, Allan Martin and Dan Madigan's collection *Digital Literacies for Learning* (2006) was the only English-language book with "digital literacies" in the title to show up in a search on Amazon.com.

The plural form fares better among English-language journal articles (e.g., Anderson & Henderson, 2004; Ba, Tally, & Tsikalas, 2002; Bawden, 2001; Doering et al., 2007; Myers, 2006; Snyder, 1999; Thomas, 2004) and conference presentations (e.g., Erstad, 2007; Lin & Lo, 2004; Steinkeuhler, 2005), however, and is now reasonably common in talk on blogs and wikis (e.g., Couros, 2007; Davies, 2007). Nonetheless, talk of digital literacy, in the singular, remains the default mode.

The authors invited to contribute to this book were chosen in light of three reasons we (the editors) identify as important grounds for promoting the idea of digital literacies in the plural. This, of course, does not mean the contributing authors would necessarily subscribe to some or all of these reasons. That was not a criterion for participating. At the same time, the positions argued by each of the contributing authors in this volume seem to us to support the case for taking the idea of digital literacies very seriously.

We believe it is important to emphasize the plurality of digital literacies because of:

- the sheer diversity of specific accounts of "digital literacy" that exist, and consequent implications of that for digital literacy policies;
- the strength and usefulness of a sociocultural perspective on literacy as practice, according to which literacy is best understood as literacies (Street, 1984; Lankshear, 1987; Gee, 1996). By extension, then, digital literacy can usefully be understood as digital *literacies*—in the plural;
- the benefits that may accrue from adopting an expansive view of digital literacies and their significance for educational learning.

A Plethora of Conceptions of Digital Literacy

As the chapters that follow attest, the most immediately obvious facts about accounts of digital literacy are that there are many of them and that there are significantly different *kinds* of concepts on offer.

David Bawden (Chapter 1) refers to Paul Gilster's (1997; Pool, 1997) claim that digital literacy involves "mastering ideas, not keystrokes." One way of distinguishing the burgeoning array of concepts of digital literacy is, indeed, to delineate those that emphasize mastery of ideas and insist on careful evaluation of information and intelligent analysis and synthesis, from those that provide lists of specific skills and techniques that are seen as necessary for qualifying as digitally literate. A second broad line of demarcation indicated by Bawden (pp. 17–32 here) involves Eshet-Alkalai's (2004) caution concerning the inconsistency between those who conceive digital literacy as "primarily concerned with technical skills, and those who see it as focused on cognitive and socio-emotional aspects of working in a digital environment."

Similarly, we might distinguish *conceptual* definitions of "digital literacy" from *"standardized operational"* definitions (Lankshear & Knobel, 2006). Conceptual definitions present views of digital literacy couched as a general idea or ideal. In one of the earliest examples of a conceptual definition Richard Lanham (1995, p. 198) claims that "literacy" has extended its semantic reach from meaning "the ability to read and write" to now meaning "the ability to understand information however presented." He emphasizes the multimediated nature of digital information and argues that to be digitally literate involves "being skilled at deciphering complex images and sounds as well as the syntactical subtleties of words." (Lanham, 1995, p. 200) Digitally literate people are "quick on [their] feet in moving from one kind of medium to another . . . know what kinds of expression fit what kinds of knowledge and become skilled at presenting [their] information in the medium that [their] audience will find easiest to understand." (ibid.) According to this ideal, digital literacy enables us to match the medium we use to the kind of information we are presenting and to the audience we are presenting it to.

Standardized operational definitions, by contrast, "operationalize" what is involved in being digitally literate in terms of certain tasks, performances, demonstrations of skills, etc., and advance these as a *standard* for general adoption. A well-known commercial variant is Certiport's Internet and Computing Core Certification (IC^3) (www.certiport.com). The website claims that "IC^3 certification helps you learn and demonstrate Internet and digital literacy through a worldwide industry standard," through training and exam certification covering Computing Fundamentals, Key Applications, and Living Online. Computing Fundamentals test items involve tasks like asking learners to click on all the "output devices" from a list containing items like joystick, monitor, speakers, keyboard, etc.; to choose among four items (one thousand, one million, one billion, one trillion) for the number of bytes in a megabyte; to create a new folder on the C drive within a simulated file manager; and to match "operating system," "application" and "utility program" to three provided definitions. The items testing Key Applications use a range of simulations and ask learners to insert content from the clipboard at the designated insertion point and exit Word without using the close box. Items assessing knowledge and skills related to Living Online use simulations to have respondents enter a subject in an email message and send the message, go to a specified address on a web page, and locate the history of sites visited in a web browser. Certiport asserts that IC^3 certification meets the technology requirements of "No Child Left Behind" legislation, with respect to ensuring that every student "regardless of . . . race, ethnicity, gender, family income, geographic location, or disability" is digitally literate by the time they finish 8th grade, and by providing "the professional development 'through electronic means' for teachers, administrators, and staff called for in No Child Left Behind's "Enhancing Education Through Technology Act."

Among the chapters that follow, those by David Bawden on origins and concepts of digital literacy (Chapter 1), Leena Rantala and Juha Suoranta on digital literacy policies in the European Union (Chapter 5), Morten Søby on digital competence with particular reference to the Norwegian context (Chapter 6), and Allan Martin on digital literacy and the digital society (Chapter 7) especially foreground the sheer diversity and complexity of conceptions of digital literacy. They situate digital literacy in relation to a web of "literacies of the digital" (Martin, Chapter 7) including ICT/computer literacy, information literacy, technological literacy, media literacy, communication literacy, visual literacy, network literacy, e-literacy, digital competence, digital Bildung, and the like. David Buckingham (Chapter 4) addresses "web literacy," "game literacy" and "writing digital media" in the context of developing an ideal of digital literacy in terms of what young people need to know about digital media. Such a larger map of concepts of digital literacy provides a lens for locating the kinds of focus represented in Genevieve Johnson's chapter on "functional internet literacy" (Chapter 2), and the chapter on "digital literacy as information savvy" by Maggie Fieldhouse and David Nicholas (Chapter 3) as contributions to developing a robust discourse of digital literacy.

This sheer variety means that digital literacy can be seen as "a framework for integrating various other literacies and skill-sets" without "the need to encompass them all" or to serve as "one literacy to rule them all" (Martin cited in Bawden, Chapter 1 here; Martin, 2006). Equally, however, it reminds us that any attempt to constitute an umbrella definition or overarching frame of digital literacy will necessarily involve reconciling the claims of myriad concepts of digital literacy, a veritable legion of digital literacies.

The Sociocultural View of Literacy as a Set of Socially Organized Practices

In the first extended English-language treatment of "digital literacy," Paul Gilster (1997, p. 1) defines digital literacy as "the ability to understand and use information in multiple formats from a wide range of sources when it is presented via computers." This, says Bawden (Chapter 1), is quite simply "literacy in the digital age . . . [T]he current form of the traditional idea of literacy *per se*—the ability to read, write, and otherwise deal with information using the technologies and formats of the time." This conception of digital literacy as what *literacy* is in the digital era opens up a second—sociocultural—line of argument for understanding "digital literacy" as a shorthand (Street 1984, p. 1) for digital *literacies*.

From a sociocultural perspective literacy is a matter of social practices (Gee, Hull & Lankshear, 1996, p. 1). Brian Street (1984, p. 1) argues that literacy "is best understood as a shorthand for the social practices and conceptions of reading and writing." Previously, Silvia Scribner and Michael Cole (1981, p. 236) had argued that literacy comprises "a set of socially organized practices which make use of a symbol system and a technology for producing and disseminating it" (see Chapter 11 here). Literacy does not simply involve knowing how to encode and decode a particular kind of script. According to Scribner and Cole it involves "applying this knowledge for specific purposes in specific contexts of use." (1981, p. 236)

This approach has two important implications for how we think about literacy so far as the plurality of digital literacies is concerned. The first is that reading (and writing) always involve particular kinds of texts and particular ways of reading (and writing) that vary enormously. The case for reading can be stated as follows:

> Whatever literacy is, it [has] something to do with *reading*. And *reading* is always *reading something*. Furthermore, if one has not *understood* [made meaning from] what one has read then one has not read it. So reading is always reading *something* with *understanding* [T]his something that one reads with understanding is always a text of a *certain type* which is read in a *certain way*. The text might be a comic book, a novel, a poem, a legal brief, a technical manual, a textbook in physics, a newspaper article, an essay in the social sciences or philosophy, a "self-help" book, a recipe, and so forth through many different types of text. Each of these different types of text requires somewhat different background knowledge and somewhat different skills. (Gee, Hull, & Lankshear, 1996, pp. 1–2).

If we extend this argument from literacy to digital literacy it involves thinking of "digital literacy" as a shorthand for the myriad social practices and conceptions of engaging in meaning making mediated by texts that are produced, received, distributed, exchanged, etc., via digital codification. Hence, to the list contained in the above quotation we may add blogs, video games, text messages, online social network pages, discussion forums, internet memes, FAQs, online search results, and so on.

Moreover, as is the case with the kinds of conventional text types previously mentioned, many types of digital texts will *themselves* take multiple forms. For example, the social practices of any two bloggers may seem as different from each other as writing an academic paper is from emailing a parent, spouse or sibling. Blogs are created and maintained for diverse purposes and as elements or dimensions of diverse social practices. These include but are far from exhausted by (combinations of) the following: as personal diaries/journals; to provide alternative accounts of events and other phenomena to those of mainstream media as part of a citizen journalist practice; to critique mainstream broadcasting of news events as part of a "news watch" affinity space; to sell products or distribute corporate news as part of commercial practice; to express personal opinions as part of one's alliance with particular points-of-view or perspectives; to archive memories (e.g., photo blogs, audio blogs, video blogs); to parody other blogs and other media; to augment fan fiction writing or drawing; to archive or index profession-related materials (e.g., hyperlinks to relevant policy documents and news reports, etc.); to augment hobbies and pastimes (e.g., collecting items, techno-gadgetry, genealogy studies, sport); to notify fans of popular culture events and information (like band tour dates, author readings and book events, art and design world developments), and so on. The sheer diversity of weblogs and weblogging practices cautions against conceiving blogging as a specific singular type.

The second implication builds further on what has just been said. It is well known that different people can read the same text in different ways and, furthermore, that some people simply cannot make sense of certain texts (despite being able to decide or encode them accurately) that other people handle with ease. Photoshopped images provide a good example here. An image of a snake pulling a cow up the side of a ravine is read by one viewer as an absolutely amazing testimony to the size and strength of a snake, and they express horror that such snakes are on the loose out there. It is read by a photoshopper as a pretty cool remix of some images to produce an absurdity that is so technically proficient it looks real. The current "LOLcats" online phenomenon (e.g., icanhascheezburger.com; www.dropline.net/cats) provides another instance. LOLcat texts typically show cats in weird poses, with captions containing strange, phonetically-spelled, syntactically odd, written language. Participating in the remixed LOLcats meme involves reading and writing distinctive language, using popular culture references, and employing certain motifs (e.g., "i can has X?"; "o hai" for "oh hello", which invokes pop culture English translations of Japanese texts; "kthnxby" for "Okay, thanks. Bye"; repeated refrains like "I is in ur Y, Xing all ur Zs," and various uses of game, computer and movie terms like "lasers on," "morph ball acquired" and "n00b," among others). Shared insider jokes about cute cats having secret lives as avid game players, as computer technicians, as having a range of magical powers, as being able to muster a range of smart weapons for different purposes, and suchlike, tap into a keen interest in the absurd often found in gaming and computer discussion boards where these kinds of images were first generated. Many of these texts appear nonsensical to "outsiders" but nonetheless answer to certain ("insider") conventions of use.

Sociocultural theorists respond to the question of how someone acquires the ability to read a particular kind of text in a particular way by emphasizing apprenticeship to social practices.

A way of reading a certain type of text is acquired *only* when it is acquired in a "fluent" or "native-like" way, by one's being embedded in (apprenticed as a member of) a *social practice* wherein people not only *read* texts of this type in these ways but also *talk* about such texts in certain ways, *hold certain beliefs and values* about them, and *socially interact* over them in certain ways . . . Texts are parts of *lived, talked, enacted, value-and-belief-laden* practices carried out in specific places and at specific times (Gee, Hull, & Lankshear, 1996, p. 3).

From a sociocultural perspective, these different ways of reading and writing and the "enculturations" that lead to becoming proficient in them are *literacies*. Engaging in these situated practices where we make meanings by relating texts to larger ways of doing and being is engaging in literacy—or, more accurately, *literacies*, since we are all apprenticed to more than one. To grasp this point is to grasp the importance of understanding that "digital literacy" must also be seen as digital literacies. Hence, when we take an expansive conception of "digital literacy," such as Gilster's, we can see that "the ability to understand and use information in multiple formats from a wide range of sources when it is presented via computers" will take diverse forms according to the many and varied social practices out of which different individuals are enabled to understand and use information and communications.

While all chapters in the book have something to say about social practices in relation to digital literacies, this is the primary role of chapters in the second half of the book (from Chapter 8 on). These chapters deal with selected aspects of digital remix, blogging, online shopping, social networking, and legal considerations that impact on digital literacies. Ola Erstad explores trajectories of remixing, looking at digital literacies from the standpoint of media production and schooling (Chapter 8). Lilia Efimova and Jonathan Grudin discuss digital literacies at work by reference to the case of employees' blogging (Chapter 9), and Julia Davies explores digital literacies of online shoppers buying and selling on eBay.com (Chapter 10). Michele Knobel and Colin Lankshear conclude the second part of the book by discussing participation in online social networking spaces in terms of digital literacy practices (Chapter 11) and by assembling and remixing some of Lawrence Lessig's work to provide a perspective on digital literacy and the law (Chapter 12).

Toward an Expansive Politics of Digital Literacy

Educational learning serves multiple ends. These include academic and scholarly ends, civic ends, personal success and fulfillment ends, and what James Paul Gee (2005; 2007, Chapter 1) calls for the good "of the soul." We would argue that during the past 50 years—and particularly during the past 25 years—the pursuit of literacy as a *sine qua non* for realizing these ends has often had counterproductive effects. A narrow focus on literacy as fluent encoding and decoding has done nothing to change familiar patterns of academic success and failure. At the same time, it has presided over escalating levels of disengagement from education that in many schools have reached crisis levels. Many souls have died or been severely damaged in the process.

> If people are to nurture their souls, they need to feel a sense of control, meaningfulness, even expertise in the face of risk and complexity. They want and need to feel like heroes in their own life stories and to feel that their stories make sense. They need to feel that they matter and that they have mattered in other people's stories. If the body feeds on food, the soul feeds on agency and meaningfulness. (Gee, 2007, p. 10)

Ironically, agency and meaningfulness are the very stuff of literacies as situated social practices. It has almost become a research cliché to cite instances of young people trapped in literacy remediation in schools whilst winning public esteem as fan fiction writers, AMV remixers, or successful gamers online. Experiences of agency and meaningfulness within learning contexts that engender it have powerful consequences for learning. Gee makes the case explicitly for video games, but it holds more widely.

> Good video games give people pleasures. These pleasures are connected to control, agency, and meaningfulness. But good games are problem-solving spaces that create deep learning, learning that is better than what we often see today in our schools. Pleasure and learning: For most people these two don't seem to go together. But that is a mistruth we have picked up at school, where we have been taught that pleasure is fun and learning is work, and, thus, that work is not fun (Gee, 2004). But, in fact, good video games are hard work and deep fun. So is good learning in other contexts. (Gee, 2007, p. 10)

What holds for video games holds in varying ways and degrees for legions of bloggers, social networkers, fanfic authors, machinima creators, photoshoppers, digital animators, music video and movie trailer exponents, who troubleshoot, collaborate, share and develop expertise, and give and receive feedback in all manner of online affinity spaces, in the process of co-learning and refining these arts in the company of others who share these affinities (Gee, 2004).

Approaching digital literacy from the standpoint of digital *literacies* can open us up to making potentially illuminating connections between literacy, learning, meaning (semantic as well as existential), and experiences of agency, efficacy, and pleasure that we might not otherwise make. The point here is not simply to import an array of digital literacies holus bolus into classrooms on the grounds that they are "engaging," or because learners who do not experience success in conventional school subject literacies can nonetheless experience success and affirmation as bloggers, gamers and fan practice aficionados—although that would be no small thing. Rather, the educational grounds for acknowledging the nature and diversity of digital literacies, and for considering where and how they might enter into *educational* learning have partly to do with the extent to which we can build bridges between learners' existing interests in these practices and more formal scholarly purposes.

In this vein Lawrence Lessig (2004, pp. 38–39; see Chapter 12 here) reports an example from a low-income area inner city Los Angeles school. In a project that involved mixing images, sound and text, led by Elizabeth Daley and Stephanie Barish, high school students with low school literacy achievement (and an open resistance to writing at school) expressed their perspectives on gun violence—with which they were very familiar. Inspired by their own video remixes, students "bumped up against the fact [that they had] to explain this . . . and really [needed] to write something". Often "they would rewrite a paragraph 5, 6, 7, 8 times, till they got it right. Because they *needed* to" (in Lessig, 2004, p. 39, our emphasis). This need was born of emotional and cognitive investment in an achievement and the will to perfect it.

The educational grounds for acknowledging the nature and diversity of everyday digital literacies and where they enter into educational learning have to do also with the extent to which we can identify principles by which digital literacies successfully recruit participants to learning and mastering them, and then translate these principles into effective approaches for pursuing bona fide educational ends (cf. Carr et al., 2006; Black, 2005, 2007; diSessa, 2000: Gee 2003, 2004, 2007; Hull, 2004; Jenkins, 2006; Lam, 2000; Shaffer, 2005).

There is a further important point to be made here concerning the plurality of literacies and the politics of literacy within formal education. The conventional *singular* educational conception of literacy as proficiency with print has done much to mask the ways language and literacy play out in formal educational settings. It is well recognized among sociocultural researchers and theorists of literacy that particular "ways with words" (Heath, 1982; 1983) are aligned consistently with experiences of academic success within scholastic settings, whereas others are aligned with educational under-achievement. This again, is practically a cliché for anyone versed in the politics and sociology of literacy. Most recently, Gee (2007) has addressed this issue in a way that has direct relevance to digital literacies.

Gee refers to an equity crisis in traditional print literacy: "poorer children do not learn to read and write as well as richer children" (Gee, 2007, p. 138). In part, this is a matter of poorer children having higher rates of functional illiteracy than richer children. More subtly, however, poorer children who become fluent encoders and decoders of alphabetic text systematically do less well in scholastic reading and writing than richer children. In the U.S. this difference is embodied in what is referred to widely as "the fourth grade slump," and educators have been aware of it for decades. This 4th grade slump names the phenomenon

> whereby many children, especially poorer children, pass early reading tests, but cannot later on in school read well enough to learn academic content. They learn early on to read, but don't know how to read to learn when they face more complex language and content as school progresses. (Gee, 2007, p. 138)

That is, literacy in the general sense of literal encoding and decoding is not the literacy that confers access to the learning that counts *scholastically* for school success. Moreover, the kinds of early language experiences that correlate with school success—with learning in content areas and not just with literacy in the sense of encoding and decoding and text-level comprehension—are not universal within societies like our own. Rather, they are more closely associated with membership of certain "primary discourses" (Gee, 1996) than others. Some children get much more early exposure than others to particular kinds of oral vocabulary and ways of talking involving complex language associated with books and school. This is language experience that prepares young people for managing language "that is 'technical' or 'specialist' or 'academic'" and not just "everyday" (Gee, 2007, p. 139). Whereas early childhood experiences that promote "phoneme awareness and home-based practice with literacy" correlate well with "success in learning to decode print" and with other dimensions of success in the early grades, these are not the best predictors of school success in 4th grade and beyond. Instead, it is getting the kinds of experience that set learners up for managing technical and specialist language that counts most (ibid.).

This is increasingly well understood, although by no means as well or widely understood as it needs to be—especially among education policy makers, education administrators and teachers. On the other hand, as researchers like Gee and a growing corpus of other scholars and authors in the learning sciences, games studies and popular culture (e.g., Johnson, 2005; Shaffer & Gee, 2005; Squire, 2008; Steinkuehler, 2008) are finding, numerous contemporary popular cultural pursuits involve highly technical and specialist styles of language. Young people across the socioeconomic spectrum engage in these practices socially with one another in informal online and offline peer learning groups. These practices include playing card games, associated video games, and interacting socially around trading card collections that tap into young children's interests in certain anime television series, and the like. They also widely involve engaging with digital artifacts of one kind or another, which entails complex vocabulary and syntax in order to understand the rules for video games, master concepts for operating specific software or technologies, to knowing how to participate effectively within online social spaces, and how to meet criteria for success in a practice or quest.

Such pursuits bestow opportunities (that come more or less *free*, with participating in them as "value adds") for achieving familiarity with particular forms of specialist and technical oral and written language. This language, however, is not necessarily *academic*—at least in the sense of academic literacy that pertains to schooling. In many contemporary popular cultural pursuits young people—as well as older people—are engaging in the kinds of language experiences that nonetheless *could* be leveraged for deep learning of an academic nature, as well as for educational learning conducive to developing competence in practical *professional* activities.

In other words, the digital literacy dimensions of these popular pursuits provide parallel forms of exposure to the kinds of language uses that some social groups have always drawn upon for scholastic success. They may not map as directly onto extant classroom practices as "middle class talk around books" does, but they *could* readily map onto a revitalized school curriculum that is developed and overseen by teachers who are experienced in leveraging learning principles and understandings from digital literacies for formal educational learning. This would involve a considerably redefined academic culture that was less about acquiring, remembering, and repeating subject content *per se*, and more about active participation in scholarly ways of doing and being (e.g., doing historical research like an historian, doing background research like a fiction writer, being a physicist or mathematician like professional physicists and mathematicians) and/or participation in professional, technical, administrative, civic, and other ways of doing and being that are germane to post-school life trajectories (cf. Gee, 2004, 2007; Gee, Hull, & Lankshear, 1996).

A good example can be found in the case of Tanaka Nanako, a 16-year-old English language learner who migrated to Canada as a non-English speaking native speaker of Mandarin Chinese. Nanako is a successful fanfiction writer who became the key informant of a three-year study by Rebecca Black (2005, 2007). When Nanako began writing online fanfiction, she had been learning English for just two and a half years. By the time the study was written up, Nanako had received over "6000 reviews of her 50 plus publicly-posted fanfiction texts" (Black 2007, p. 120). While a somewhat atypical case, this kind of success makes Nanako a good example of how engaging in fanfiction writing among peers can, over time, contribute to young people becoming accomplished narrative writers.

Black describes how Nanako's "author notes" to readers at the start, middle, or end of her fanfic chapters initially apologized for grammatical and spelling errors in the fictions, and how these evolved into seeking specific feedback from reviewers with respect to English grammar and spelling, and plot development. Black found that Nanako explicitly incorporated reviewer feedback into subsequent chapter revisions (cf., Black, 2005, p. 123). She argues that while Nanako's English-language development was supported in school, reviewer feedback on grammar, spelling, and such in her fanfiction also contributed directly to enhancing Nanako's English writing proficiency. Furthermore, Nanako explained in an interview with Black (2006) that she had come to realize that many of her schoolmates "were largely unaware of either Chinese or Japanese history" and that the same might hold for the readers of her fanfiction as well. Nanako had decided to focus more on the "rich histories of these two countries" (Black, 2006, p. 16) and had produced two fanfics; one that combined elements of the movie, *Memories of a Geisha*, and the anime character, Sakura (from the *Card Captor Sakura* series), and another "set in 1910 Kyoto, Japan, [which] centers on Sakura's struggles with an arranged marriage" (ibid.). Black describes how Nanako also plans to "compose a historical fiction based on the second Sino-Japanese war, or the war fought between China and Japan from 1937–1945" (ibid.). Nanako explained that "her process of writing such texts is also an opportunity for her to 'learn more about [her] own culture and history' because she often must do research to effectively represent the social and historical details in her fictions" (Black, 2006, p. 16). Such authorial dispositions, processes, and commitments to polished writing are very much valued in schools and beyond, and are practiced as a matter of course within fanfiction affinity spaces.

Furthermore, as Gee argues, participating in digital literacy practices like gaming, machinima, digital animating, fanfiction writing, blogging and the like, provides opportunities for gaining *situated* rather than merely verbal (or literal) meanings for concepts, processes and functions. Situated rather than literal meanings are, precisely, the kinds of meanings that underpin deep understanding and competence, whether in work practices or academic disciplines. They mark the difference between merely being able to parrot back content (which may be good enough for passing school tests, but not for performing with distinction in real world tasks) and attaining sound theoretical understandings and being able to apply these in concrete practical settings (displaying competence).

Along with valuable legacies of engagement with complex technical and specialized language, and immersion in situated meaning making, engaging in digital literacies like gaming, computer modeling, simulations, and popular culture-creating within activities like machinima making, Anime Music Video making, and the like, can lead to developing

> a productive reflective stance on design (including content) and to the formation of tech-savvy identities, both of which "are particularly important for today's high tech world." [Crucially, however,] these things don't just happen all by themselves. They require guidance, in one form or another, from adults and more masterful peers. (Gee, 2007, p. 138)

Gee raises two issues that go deep to the heart of the rationale for this book and that bespeak the wisdom of taking an expansive approach to digital literacies.

First, and as we might reasonably expect, early evidence (e.g., Neuman & Celano, 2006) indicates that we are already witnessing the emergence of a structural digital literacy inequity along the lines of richer children-poorer children alongside the traditional literacy gap. In this event, "richer children [will] attain productive stances toward design and tech-savvy identities to a greater degree than poorer ones" (Gee, 2007, p. 138), thereby creating a new equity gap involving skills and identities that may be crucially tied to success in the contemporary world.

> [E]vidence is beginning to show that just giving young people access to technologies is not enough. They need—just as they do for books—adult mentoring and rich learning systems built around the technologies, otherwise the full potential of these technologies is not realized for these children (Gee, 2007, p. 138).

Second, the distinctive socio-technical accompaniments of digital literacies—the myriad "learning incidentals" that come free with the online and offline learning systems attaching to digital literacy practices within affinity spaces of any kind, but including popular cultural forms—suggest the possibility of addressing "the new gap (the tech-savvy gap) in such a way that we [simultaneously] address the old gap, the gap in regard to traditional print-based literacy" (ibid.).

Approaching digital literacy in terms of "digital literacies" allows for the kinds of analysis of social practices that identify key points at which effective learning is triggered within efficient socio-technical learning systems as well as key learning principles that can be adapted and leveraged for equitable educational learning. Taking an expansive view of digital literacies—one that includes popular cultural practices, everyday practices like workplace blogging, online shopping and participation in online network sites—extends the scope for identifying and understanding points at which these same conducive processes and principles operate within digital literacies that are increasingly part of the everyday lives of educators at large.

Conclusion

We began by saying that the authors invited to contribute to this volume were chosen on the basis of the excellent contributions we thought they could in various ways make to (i) demonstrating the kind of diversity that exists among concepts of digital literacy; (ii) modeling the strengths and usefulness of a sociocultural approach to understanding digital literacy as a plural phenomenon comprising many digital literacies; and (iii) establishing the benefits of adopting an expansive view of digital literacies and their significance for educational learning. We believe they have done exactly that, and trust that readers will share this assessment as they explore the chapters that follow and the rich tapestry of perspectives on digital literacy that they provide.

References

Anderson N. & Henderson, M. (2004). Blended models of sustaining teacher professional development in digital literacies. *E-Learning*, 1(3), 383–394.

Ba, H., Tally, W., & Tsikalas, K. (2002). Investigating children's emerging digital literacies. *Journal of Technology, Learning and Assessment*. 1(4). Available at: http://www.jtla.org

Bawden, D. (2001). Information and digital literacies: a review of concepts. *Journal of Documen-*

tation. 57(2), 218–259.

Black, R. (2005). Access and affiliation: The literacy and composition practices of English language learners in an online fanfiction community. *Journal of Adolescent & Adult Literacy. 49* (2), 118–128.

———. (2006). Not Just the OMG standard: Reader feedback and language, literacy, and culture in online fanfiction. Paper presented to the American Educational Research Association annual conference. San Francisco, April.

———. (2007). Digital design: English language learners and reader reviews in online fanfiction. In M. Knobel & C. Lankshear (Eds.), *A new literacies sampler* (pp. 115–136). New York: Peter Lang.

Carr, D., Buckingham, D., Burn, A., & Schott, G. (2006). *Computer games: Text, narrative and play.* Malde, MA: Polity Press.

Couros, A. (2007). Digital Literacies & Emerging Educational Technologies—A Wiki. Available at: http://couros.wikispaces.com/emerging+technologies

Davies, J. (2007). Digital Literacies Blog. Available at: http://digital-literacies.blogspot.com

diSessa, A. (2000). *Changing minds: Computers, learning, and literacy.* Cambridge, Mass.: MIT Press.

Doering, A., et al. (2007). Infusing multimodal tools and digital literacies into an English education program. *English Education.* 40(1), 41–60.

Erstad, O. (2007). Polarities and potentials—digital literacies in domains of science. Paper presented to the New Literacies: conflict or confluence? Conference, The Danish University of Education, Copenhagen, 19 June.

Eshet-Alkalai, Y. (2004), Digital literacy: a conceptual framework for survival skills in the digital era, *Journal of Educational Multimedia and Hypermedia,* 139(1), 93–106. Available at: http://www.openu.ac.il/Personal_sites/download/Digital-literacy2004-JEMH.pdf

Gee, J. P. (1996) *Social linguistics and literacies: Ideology in discourses.* 2nd edition. London: Taylor & Francis.

———. (2003). *What video games have to teach us about learning and literacy.* New York: Palgrave/ Macmillan.

———. (2004). *Situated language and learning: A Critique of traditional schooling.* London: Routledge.

———. (2005). *Why video games are good for your soul: Pleasure and learning.* Melbourne: Common Ground.

———. (2007). *Good video games + good learning: collected essays on video games, learning and literacy.* New York: Peter Lang.

Gee, J. P., Hull, G. and Lankshear, C. (1996/2003) *The new work order: behind the language of the new capitalism.* Boulder, CO: Westview Press.

Gilster, P. (1997) *Digital literacy.* New York: John Wiley & Sons Inc.

Heath, S. (1982). What no bedtime story means: Narrative skills at home and school. *Language in Society* 11(1), 49–76.

———. (1983). *Ways with words: language, life and work in community and classrooms.* Cambridge University Press, Cambridge.

Hull, G. (2004). Youth culture and digital media: New literacies for new times. *Research in the Teaching of English,* 38(2), 229–233.

Jenkins, H. (with Purushotma, R., Clinton, K., Weigel, M., & Robison, A.) (2006). *Confronting the challenges of participatory culture: media education for the 21ˢᵗ century*. Chicago, IL: MacArthur Foundation. Available at: http://homeinc.org/pdf/2007_MLC_Program.pdf

Johnson, S. (2005). *Everything Bad Is Good for You: How Today's Popular Culture Is Actually Making Us Smarter*. New York: Riverhead Books.

Lam, Wan Shun Eva (2000). Second language literacy and the design of the self: a case study of a teenager writing on the internet. *TESOL Quarterly*, 34 (3), 457–483.

Lanham, R. (1995). Digital literacy, *Scientific American*, 273(3), 160–161.

Lankshear, C. (1987). *Literacy, Schooling and Revolution*. London: Falmer Press.

Lankshear, C. and Knobel, M. (2006). Digital literacies: policy, pedagogy and research considerations for education. *Digital Kompetanse: Nordic Journal of Digital Literacy*, 1(1), 12–24.

Lessig, L. (2004). *Free culture: How big media uses technology and the law to lock down culture and control creativity*. New York: Penguin.

Lin, A. & Lo, J. (2004). New youth digital literacies and mobile connectivity: Text-messaging among Hong Kong college students. Paper presented to the International Conference on Mobile Communication and Social Change, Seoul, South Korea. October.

Martin, A. (2006). Literacies for the digital age, in A. Martin and D. Madigan (Eds.), *Digital literacies for learning* (pp. 3–25). London: Facet Publishing.

Martin, A. & Madigan, D. (Eds.) (2006). *Digital literacies for learning*. London. Facet Publishing.

Myers, J. (2006). Literacy practices and digital literacies: A Commentary on Swenson, Rozema, Young, McGrail, and Whitin [Discussion of Beliefs About Technology and the Preparation of English Teachers: Beginning the Conversation by Janet Swenson and others]. *Contemporary Issues in Technology and Teacher Education*. 6(1), 61–66.

Neuman, S. B. & Celano, D. (2006). The knowledge gap: Implications of leveling the playing field for low-income and middle-income children. *Reading Research Quarterly*, 41(2), 176–201.

Pool, C. (1997). A conversation with Paul Gilster. *Educational Leadership*. 55, 6–11.

Scribner, S. & Cole, M. (1981) *The psychology of literacy*. Cambridge, MA: Harvard University Press.

Shaffer, D. (2007). *How computer games help children learn*. New York: Palgrave/Macmillan.

Shaffer, D. & Gee, J. (2005). Before every child is left behind: How epistemic games can solve the coming crisis in education. Madison: University of Wisconsin-Madison. Available at http://www.academiccolab.org/resources/documents/learning_crisis.pdf

Snyder, I. (1999). Digital literacies: renegotiating the visual and the verbal in communication. *Prospect*. 14(3), 13–23.

Squire, K. (2008). Video game literacy: a literacy of expertise. In J. Coiro, M. Knobel, C. Lankshear and D. Leu (Eds.), *Handbook of research on new literacies*. Mahwah, NJ: Lawrence Erlbaum Associates/Taylor and Francis-Routledge.

Steinkeuhler, C. (2005). Digital literacies and massively multiplayer online games. Paper presented to the National Reading Conference (NRC), Miami, FL, November.

———. (2008). Cognition and literacy in massively multiplayer online games. In J. Coiro, M. Knobel, C. Lankshear & D. Leu (Eds.), *Handbook of research on new literacies*. Mahwah, NJ: Lawrence Erlbaum Associates/Taylor & Francis-Routledge.

Street, B. (1984). *Literacy in theory and practice*. Cambridge: Cambridge University Press.

Thomas, A. (2004). Digital literacies of the cybergirl. *E-Learning*, 1(3), 358–382.

CHAPTER ONE

Origins and Concepts of Digital Literacy

DAVID BAWDEN

Introduction

The purpose of this chapter is to describe the emergence and development of the idea of "digital literacy" and to show how it relates to the various other "literacies of information." This is a topic whose terminology is very confused. Among those authors who have tried to disentangle it are Bawden (2001), Bawden and Robinson (2002), Kope (2006), Martin (2006a, 2006b), and Williams and Minnian (2007). Not only must the idea of digital literacy find its place among information literacy, computer literacy, ICT literacy, e-literacy, network literacy, and media literacy, but it must also be matched against terms which avoid the "literacy" idea, such as informacy and information fluency. Indeed in some cases, mention of information or anything similar is avoided—particularly in workplace settings—as in "basic skills," "Internet savvy," or "smart working" (Robinson et al., 2005).

The Concept of Digital Literacy

The concept of digital literacy, as the term is now generally used, was introduced by Paul Gilster, in his book of the same name (Gilster, 1997). As will be seen later, Gilster did not provide lists of skills, competences or attitudes defining what it is to be digitally literate. Rather, he explained it quite generally, as an ability to understand and to use information from a variety of digital sources and regarded it simply as literacy in the digital age. It is therefore the current form of the traditional idea of literacy *per se*—the ability to read, write and otherwise deal with information using the technologies and formats of the time—and an essential life skill. This generic expression of the idea, although it has irritated some commentators, is one of the strengths of Gilster's concept, allowing it to be applied without concern for the sometimes restrictive "competence lists" which have afflicted some other descriptions of the literacies of information.

Gilster was not the first to use the phrase "digital literacy;" it had been applied throughout the 1990s by a number of authors, who used it to mean essentially an ability to read and comprehend information items in the hypertext or multimedia formats which were then becoming available (Bawden, 2001). Typical of these is Lanham (1995), who regarded it as a kind of "multimedia literacy," quite different from traditional literacy. His argument was that since a digital source could generate many forms of information—text, images, sounds, etc.—a new form of literacy was necessary, in order to make sense of these new forms of presentation. While this is certainly an important aspect of the wider concept of digital literacy, it is too restrictive, and arguably too much influenced by the technology of its times, to be of as much lasting value as Gilster's broader conception. Several conceptions of this kind are reviewed by Eshet (2002), who concludes, like Gilster, that digital literacy must be more than the ability to use digital sources effectively; it is a special kind of mindset or thinking.

In his 1997 book, Gilster states this explicitly—"digital literacy is about mastering ideas, not keystrokes"—thus distinguishing his conception from the more limited "technical skills" view of digital literacy. It is, he says, "cognition of what you see on the computer screen when you use a networked medium. It places demands upon you that were always present, though less visible, in the analog media of newspapers and TV. At the same time, it conjures up a new set of challenges that require you to approach networked computers without preconceptions. Not only must you acquire the skill of finding things, you must also acquire the ability to use these things in your life" (pp. 1–2).

The mention of "networked computers" is a reminder that Gilster's book was written at the time of the first flush of enthusiasm for the internet, and many of his examples and instances are internet-related. In his introduction to the book, he sets the challenge of effective use of the internet into the long sequence of adaptation to new information technologies beginning with the clay tablets of the Sumerian period: "technology demands of us, as it did of them, a sense of possibilities, and a willingness to adapt our skills to an evocative new medium. And that is the heart of information literacy. Our experience of the Internet will be determined by how we master its core competencies." The casual reader might assume, as did some reviewers of the book, that Gilster's digital literacy and effective use of the internet were essentially the same.

This is by no means the case. Gilster states explicitly that "no-one is asking you to give up other sources of information just to use the Internet," that "the Internet should be considered one among many sources of ideas in a technological society" and that evidence must be gathered from many sources, not just the world wide web, for the task of "knowledge assembly." More than this, although he gives, as perhaps the single clearest explanation in the book, the idea that digital literacy is "the ability to understand and use information in multiple formats from a wide variety of sources when it is presented via computers," he allows that there are non-digital formats as well. He specifically noted that digital literacy involved an understanding of how to complement digital resources with such things as reference works in libraries, printed newspapers and magazines, radio and television, and printed works of literature, expressing a particular fondness for the last. While the inexorable shift to digital formats in the decade since his book appeared might make these qualifications and caveats seem less important than when they were written, it is important to note that from its first mention, Gilster's digital literacy is not about any particular technology, not even—paradoxically, given the term—digital technology itself. It is about the ideas and mindsets, within which particular skills and competences operate, and about information and information resources, in whatever format. The term itself is quite reasonable in this context, since all information today is either digital, has been digital, or could be digital.

Gilster's book does not give a particularly clear and coherent account of digital literacy itself, or of the skills and attitudes that underlie it; it is, rather, an impressionistic and wide-ranging account, which may lead to some confusion for anyone attempting to express the ideas within a structured framework, and to determine which are of primary importance; see Bawden (2001) for a discussion. Reviewers of Gilster's book were somewhat critical of this aspect, describing it as, for example, "not organised very well or very logically" (Nicholas & Williams, 1998) and noting that "useful information for the reader is scattered in bits and pieces" (Bunz, 1997).

Although there is nowhere in the book any specified list of skills, competences, etc. associated with the general idea of digital literacy, a list may be derived from the text (Bawden, 2001). In brief, this includes:

- "knowledge assembly," building a "reliable information hoard" from diverse sources
- retrieval skills, plus "critical thinking" for making informed judgements about retrieved information, with wariness about the validity and completeness of internet sources
- reading and understanding non-sequential and dynamic material
- awareness of the value of traditional tools in conjunction with networked media
- awareness of "people networks" as sources of advice and help
- using filters and agents to manage incoming information
- being comfortable with publishing and communicating information, as well as accessing it

Gilster summarizes these at one point in the book by suggesting that there are four core competencies of digital literacy: Internet searching, hypertext navigation, knowledge assembly, and content evaluation. This list however seems to miss out some of the issues quoted at various places as significant.

Another aspect of the somewhat informal nature of the book's material is that there is no clear statement of whether any of the various aspects is central, fundamental or most important (Bawden, 2001). At various points, content evaluation and critical thinking is referred to as "most essential," "most significant" and "overarching." At other points, the ability to read and understand dynamic non-sequential information is cited as the basis for the concept. In still other sections it is the finding of information from various sources which is given priority.

This concept of digital literacy is plainly a very broad span, from specific skills and competences to rather general awareness and perspective. Developments in the decade since it was proposed, from the ubiquity of Google to the rise of social networking have validated the list as representing, in broad terms, the needed form of literacy for the present time.

Gilster's book, and the ideas in it, achieved relatively little impact in the years following its publication. Whether this was because of its idiosyncratic writing style, the fact that it appeared as a paperback, and reasonably "popular," book, rather than a journal article, or simply that the phrase "digital literacy" denoted—to those who had not read it—an exclusively technical approach, it is difficult to say.

Origins: Information and Computer Literacies

Gilster's idea of digital literacy did not appear "out of the blue." There was already a substantial set of literature and practical experience around the ideas of information literacy and computer literacy: for detailed accounts of the early history of these ideas; see Bawden (2001), Snavely and Cooper (1997), and Behrens (1994); for accounts of later developments, see Andretta (2005, 2007), Virkus (2003), and Webber and Johnson (2000).

Both of these terms (together with equivalents such as "IT literacy") originated largely to describe sets of specific skills and competences needed for finding and handling information in computerized form. "Computer literacy" was the term mainly in vogue through the 1980s, with "information literacy" gaining popularity in the 1990s. The former term, still in use in some quarters, has for the most part retained its original and straightforward "skill set" implication, based on being able to operate commonly used software packages effectively. The latter has broadened its meaning, has been accepted as a multifaceted concept, and has been understood in various ways. The information literacy concept has been largely, though not exclusively, promoted by the academic library community. It slowly grew to take on a wider meaning than its original skills-based formulation, and started to encompass aspects of the evaluation of information, and an appreciation of the nature of information resources. Though still focused on computerized information, which was believed to be most problematic to its users, it grew to encompass the use of printed resources, and hence to overlap with such concepts as "library literacy" and "media literacy" (Bawden, 2001).

At a relatively early stage in the development of the concept, in 1989, the American Library Association promulgated a six-stage model for information literacy, which has had great influence. This regarded information literacy as comprising six aspects of a linear process of information handling:

- recognizing a need for information
- identifying what information is needed
- finding the information

- evaluating the information
- organizing the information
- using the information

This still forms the basis for most approaches to information literacy to the present day, though much elaborated, extended, and refined, and with numerous variants differing in detail and emphasis. Usually this involves adding extra aspects, e.g., splitting "finding information" into "choosing resources" and "searching" and "accessing the items identified," or adding aspects such as "communicating information," or "storing / archiving / deleting information," where they are important in a particular context. An example is the "seven pillars" model, developed by SCONUL (Society of College, National, and University Libraries) in the UK (SCONUL, 2006), which distinguishes the following seven aspects:

- recognize information need
- distinguish ways of addressing gap
- construct strategies for locating
- locate and access
- compare and evaluate
- organize, apply and communicate
- synthesize and create

This understanding of information literacy goes somewhat beyond the skills-based computer literacy model, by including softer skills such as evaluation of information and recognition of information need, but is still a rather prescriptive and formulaic approach, based upon the assumption of a formally expressed information need. It is also very much a model used for planning training courses in information literacy, and widely used for that purpose within academic libraries, also forming the basis for interactive tutorials.

During the 1990s, an alternative viewpoint emerged, although it never challenged the popularity of the "six stages" style of model. This viewpoint saw information literacy less as a series of competences to be mastered and more as a set of general knowledge and attitudes to be possessed by an information literate person. Notable is the set of seven key characteristics presented by Bruce (1994, 1997), such that the information literate person is one who:

- engages in independent self-directed learning
- uses information processes

- uses a variety of information technologies and systems
- has internalized values that promote information use
- has a sound knowledge of the world of information
- approaches information critically
- has a personal information style

An even broader approach is that of Shapiro and Hughes (1996), who envisaged a concept of, and curriculum for, a kind of computer literacy comprising seven components:

- tool literacy—competence in using hardware and software tools
- resource literacy—understanding forms of, and access to, information resources
- social-structural literacy—understanding the production and social significance of information
- research literacy—using IT tools for research and scholarship
- publishing literacy—ability to communicate and publish information
- emerging technologies literacy—understanding of new developments in IT
- critical literacy—ability to evaluate the benefits of new technologies (note this is not the same as "critical thinking," which is often regarded as a component of information literacy)

Somewhat similar broad concepts, combining general knowledge and attitudes with specific skills, have also been described under the headings of "network literacy" (McClure, 1994), "informacy" (Neelameghan, 1995), and "mediacy" (Inoue, Naito & Koshizuka, 1997). For comparisons, see Bawden (2001) and Bawden and Robinson (2002), but, in essence: the first focuses on digital information in networked form, and is synonymous with "Internet literacy"; the second implies traditional literacy, plus information literacy; while the third emphasizes an ability to deal with digital information in a variety of media.

It seems clear that Gilster's digital literacy is to be located among these proposals; as a very broad concept, not restricted to any particular technology or form or information, and focusing on personal capabilities and attributes, rather than on any particular skill set. Its advantage over the others is its combination of the specific and the general, and (perhaps ironically) its lack of a strong structure, so that it is a general concept adaptable to changing times and concerns. What Gilster wrote with examples of search engines, databases and mailing lists works just as well with examples of folksonomies and social media, social networking sites, and weblogs. The principles outlast the specific systems and technologies.

Developing the Theme

As noted above, for most of the decade following the publication of Gilster's book, the concept of digital literacy received relatively little attention, compared with the enthusiasm for the more prescriptively defined "information literacy," used as the basis for many training programs and tutorials, particularly in higher education. Some attempts were made to derive specific lists of competences from Gilster's conception for use in training programs (Bawden, 2001), but these seem a somewhat inappropriate development, and have not gained wide interest.

Continuing confusion of terminology makes the development and use of the concept difficult to follow. Eshet-Alkalai (2004) suggests that "indistinct use of the term causes ambiguity, and leads to misunderstanding, misconceptions, and poor communication" and that there is a particular inconsistency between those who regard digital literacy as primarily concerned with technical skills and those who see it as focused on cognitive and socio-emotional aspects of working in a digital environment.

While some commentators during this period have used the digital literacy terminology in Gilster's sense—a broad concept with its emphasis on knowledge assembly from diverse sources and on critical thinking—some have still equated it with computer literacy, focusing on IT skills, as part of a wider information literacy (see, for example, Williams & Minnian, 2007), while others have equated it with network literacy, focusing on effective use of internet and other networked resources (see, for example, Hargittai, 2005, and Kauhanen-Simanainen, 2007). Burniske (2007) uses it for a concept very much focused on the "critical thinking" aspect, including: the critical and tactful use of language; the critical evaluation of websites; the analysis of visual content on the web; the analysis of digital information for credibility, logic and embedded emotional content; and the practice of good ethics and etiquette on the internet. Other uses of the term are noted by Eshet-Alkalai (2004).

To add to the confusion, other terms have been used for what appears to be very much Gilster's idea of digital literacy. The phrase "e-literacy," stemming from "electronic literacy," and still generally used as a synonym for skills-based computer literacy, has been adopted in some quarters as virtually synonymous with digital literacy, as in the definition in a Leeds University (UK) glossary of teaching technology:

> e-literacy—not to be confused with illiteracy, e-literacy is a much debated topic which goes some way to combine the traditional skills of computer literacy, aspects of information literacy (the ability to find, organize and make use of digital information) with issues of interpretation, knowledge construction and expression (http://www.leeds.ac.uk/glossaries)

It has been seriously suggested, as implied in the above definition, that a main reason that the e-literacy terminology has not been widely adopted is because of the potential confusion with illiteracy in spoken discourse; at all events, the definition above shows a close link with Gilster's conception. Martin (2003, 2005) similarly presents e-literacy as a central concept, drawing on a range of other literacies—information, media, computer/ICT and even "moral literacy"—and involving awareness, understanding and reflective evaluation as well as skills—very similar to Gilster. Indeed, Martin (2006b) suggests that digital literacy and e-literacy are synonymous. Kope (2006) also reviews the concept, arguing that it should be understood as having a component of "academic literacy" close to the "research literacy," learning styles, interpretation and integration of earlier writers.

Conversely, the "information literacy" terminology is still used for concepts seemingly very close to Gilster's. An example is a training program in "information and critical literacies," which offers a non-linear adaption of the traditional linear model of information literacy instruction (Markless & Streatfield, 2007). This has three inter-linked elements:

- connecting with information (orientation, exploring, focusing, locating)
- interacting with information (thinking critically, evaluating)
- making use of information (transforming, communicating, applying)

With its non-linear structure, and emphasis on critical thinking and communication, this seems very similar to Gilster's digital literacy, despite the alternative choice of name.

The more general digital literacy concept, with specific recognition of Gilster's concept as its basis, was used as the basis for a two-week professional development course for library / information specialists from Central/Eastern Europe and Central Asia at the Central European University in Budapest, from 1997 to 2001 (Bawden & Robinson, 2002). The course initially focused very much on skills and competences for effective use of the internet (Robinson et. al., 2000), but—as participants year-on-year came with greater familiarity with this—it changed focus to consider more general aspects of the use of information sources generally and networked information in particular. Gilster's digital literacy was used explicitly as the unifying theme. In the context of the new countries and emergent democracies of this region, the idea of digital literacy, and particularly its critical thinking component, proved to be a valuable focus for structuring the course. Indeed, the promotion of critical thinking within a digital literacy framework has been put forward as one of the principles that underlie the role of libraries and other information providers in supporting open societies (Robinson & Bawden, 2001).

A renewed interest in Gilster's digital literacy ideas, ten years on from their original publication, may be seen, most notably in an edited book with the phrase in its title, and with a chapter contributed by Gilster (Martin & Madigan, 2006a). [Though the publication date of the book suggests an earlier appearance, it appeared almost exactly a decade after Gilster's original.] The preface (Martin & Madigan, 2006b) acknowledges the significance of Gilster's concept a decade on—in a world in which networked information has expanded into all aspects of life—and in particular the importance of "ideas, not keystrokes" at its basis.

The book's overall theme is summed up by saying "Digital literacy may have some merit as an integrating (but not overarching) concept that focuses upon the digital without limiting itself to computer skills and which comes with little historical baggage" (Martin, 2006a). This seems a reasonable enough assessment of the status of the concept, ten years on. Any view of information and its use that did not focus upon the digital would be perverse at the present time, while the lack of "historical baggage"—arguments about the meaning of terms, assessments of whether they have positive or negative connotations, and turf wars as to which community can lay claim to them—is a definite advantage. The "integrative" aspect also silences many unproductive arguments. Digital literacy touches on and includes many things that it does not claim to own. It encompasses the presentation of information, without subsuming creative writing and visualization. It encompasses the evaluation of information, without claiming systematic reviewing and meta-analysis as its own. It includes organization of information but lays no claim to the construction and operation of terminologies, taxonomies and thesauri. And so on.

Gilster (2006) uses his contribution to this volume not to revisit his original ideas but rather to draw attention to one specific development of the intervening decade, and its implications. This is the increasing overlap between "content" (formally published materials) and "communication" (informal messages); the former is represented by books, journal articles, etc., the latter by letters, diary entries, etc. Distinct entities in printed media, they overlap in the digital realm through such things as blogs and wikis, undeniably communication, but with the potential to generate content. This has implications not just for the day-to-day practices of scholars, and the activities of librarians, but also for the meaning of concepts such as "collection." Navigating the products of this "digital fusion," with new products and forms of information always likely to emerge is, for Gilster, the major current challenge. Although he does not say so in this chapter, this seems a logical extension of his earlier vision, involving as it does ideas of search and navigation, knowledge of resources, and knowledge assembly.

The digital literacy concept has also been central to the DigEuLit project, which took a "Gilster-like" broad approach in defining digital literacy as:

> the awareness, attitude and ability of individuals to appropriately use digital tools and facilities to identify, access, manage, integrate, evaluate, analyse and synthesise digital resources, construct new knowledge, create media expressions, and communicate with others, in the context of specific life situations, in order to enable constructive social action; and to reflect upon this process. (Martin, 2006b)

This is extended into a description of thirteen specific processes (e.g., evaluation, synthesis, reflection) drawn from this definition, rather in the manner of the linear information literacy models defined above. Distinguishing digital literacy from these, Martin (2006b) notes that it is broader than information literacy, ICT literacy, etc., and subsumes a number of these individual literacies. He notes that it is also a quality that will vary according to each individual's life circumstances, and will change and develop over time, since it involves attitudes and personal qualities as well as knowledge and skills. Like Gilster, he sees it as a life skill, not particularly associated with formal education.

In rather similar vein, Eshet-Alkalai (2004) describes a new conceptual model for digital literacy, as a "survival skill in the digital era," though largely derived from, and mainly applicable to, the context of formal education. It is based on an integration of five other "literacies": photo-visual literacy (the understanding of visual representations); reproduction literacy (creative re-use of existing materials); information literacy (understood as largely concerned with the evaluation of information); branching literacy (essentially the ability to read and understand hypermedia); and social-emotional literacy (behaving correctly and sensibly in cyberspace). This appears to have much in common with the ideas of Gilster and Martin.

Finally, we should notice that what is commonly taken as the central theme of digital literacy—an ability to synthesize and integrate information from varied sources—is gaining increased notice as a crucial ability, in areas of study quite remote from those in which digital literacy is discussed. A striking example is Howard Gardner's concept of the "synthesizing mind," which was identified as a breakthrough idea in 2006 by the *Harvard Business Review* (Gardner, 2006).

Understanding Digital Literacy Today

Despite some continuing inconsistency in the use of the term, we see that several authors, following Paul Gilster, are using "digital literacy" to denote a broad concept, linking together other relevant literacies, based on computer/ICT competences and skills, but focused on "softer" skills of information evaluation and knowledge assembly, together with a set of understandings and attitudes.

This is also referred to by other names, particularly e-literacy and, by some, information literacy. However, the former has not gained popularity, while it is strongly associated with the linear models espoused by the library community. Digital literacy seems an appropriate and sensible name, in an age where information comes mainly in this form; though with the caveat that an important part of digital literacy is knowing when to use a non-digital source.

Digital literacy in this sense is a framework for integrating various other literacies and skill-sets, though it does not need to encompass them all; as Martin (2006a) puts it, we do not need "one literacy to rule them all." And, while it may be possible to produce lists of the components of digital literacy, and to show how they fit together, it is not sensible to try to reduce it to a finite number of linear stages. Nor is it sensible to suggest that one specific model of digital literacy will be appropriate for all people or, indeed, for one person over all their lifetime. Updating of understanding and competence will be necessary, as individual circumstances change, and as changes in the digital information environment bring the need for new fresh understanding and new competences; as Martin (2006a) puts it, digital literacy is "a condition, not a threshold."

With these caveats, we might set out the four generally agreed components of digital literacy, as they emerge from the authors quoted above, in this way:

1. underpinnings
   - literacy *per se*
   - Computer / ICT literacy

These "underpinnings" reflect the rather traditional skills, of which we may now need to regard computer literacy as one, which make up an older idea of literacy, and an ability to function in society. It seems an open question as to whether they should be regarded as a part of digital literacy (perhaps in its formulation as "smart working" or "basic skills") or whether they should be assumed, before digital literacy is grafted on.

2. background knowledge
   - the world of information
   - nature of information resources

This is the kind of knowledge that was assumed of any educated person, in the days when information came as books, newspapers and magazines, academic journals, professional reports, and not much else, and was largely accessed through physical print-on-paper libraries. The well-understood "publication chain"—from author to archivist, passing through editors, publishers, booksellers, librarians and the rest—lasted as a sensible concept well into the computer age. Now, it is largely meaningless, and there is no clear model to replace it. Nonetheless, attaining as good an understanding of what the new forms of information are, and where they fit into the world of digital information, has to be an essential start in being digitally literate.

3. central competencies
   - reading and understanding digital and non-digital formats
   - creating and communicating digital information
   - evaluation of information
   - knowledge assembly
   - information literacy
   - media literacy

These are the basic skills and competences, without which any claim to digital literacy has to be regarded skeptically. They are a remarkably wide set, and it would be sobering to try to assess to what degree they are possessed in the various countries of the world.

4. attitudes and perspectives
   - independent learning
   - moral / social literacy

These attitudes and perspectives are perhaps what make the link between the new concept of digital literacy, and an older idea of literacy, in vogue over two hundred years ago. It is not enough to have skills and competences, they must be grounded in some moral framework, strongly associated with being an educated, or as our ancestors would have said, a "lettered," person. They are arguably the most difficult to teach or inculcate of all the components, but they come closest to living up to the meaning of information from "infomare"; the transforming, structuring force.

Taken as a whole, we see that the "underpinnings" give the basic skill sets without which little can be achieved. The "background knowledge" complements them, by giving the necessary understanding of the way in which digital and non-digital information is created and communicated, and of the various forms of resources which result. The competencies are essentially those proposed by Gilster, phrased in the terms of later authors. "Information literacy" implies competences in actively finding and using information in "pull" mode, while "media literacy" implies an ability to deal with information formats "pushed" at the user. Finally, the attitudes and perspectives reflect the idea that the ultimate purpose of digital literacy is to help each person learn what is necessary for their particular situation. "Moral / social literacy" reflects the need for an understanding of sensible and correct behavior in the digital environment and may include issues of privacy and security.

At the heart of this conception are ideas of understanding, meaning, and context (Bawden, 2001; Pilerot, 2006), following Gilster's "ideas, not keystrokes." It does not seem unreasonable to regard this kind of literacy, expressed appropriately according to the context, as an essential requirement for life in a digital age.

References

Andretta, S. (Ed.). (2007). *Change and challenge: Information literacy for the 21ˢᵗ century.* Adelaide: Auslib Press.

———. (2005). *Information literacy: A practitioner's guide.* Oxford: Chandos Publishing.

Bawden, D. (2001). Information and digital literacies: a review of concepts. *Journal of Documentation, 57(2),* 218–259.

Bawden, D. & Robinson, L. (2002). Promoting literacy in a digital age: approaches to training for information literacy, *Learned Publishing, 15(4),* 297–301, Retrieved November 30, 2007, from http://www.ingenta.com/content/alpsp/lp/2002

Behrens, S. (1994). A conceptual analysis and historical overview of information literacy. *College and Research Libraries, 55(4),* 309–322.

Bruce, C. (1997). *The seven faces of information literacy.* Adelaide: Auslib Press.

———. (1994). Portrait of an information literate person, *HERDSA News, 16(3),* 9–11.

Bunz, U. (1997). Many words do not equal much content, *Computer-Mediated Communication Magazine,* Retrieved November 30, 2007, from http://www.december.com/cmc/mag/1997/oct/bunz.html

Burniske, R.W. (2007). *Literacy in the digital age* (2nd ed.). Thousand Oaks, CA: Corwin Press.

Eshet, Y. (2002). Digital literacy: A new terminology framework and its application to the design of meaningful technology-based learning environments, In P. Barker and S. Rebelsky (Eds.), *Proceedings of the World Conference on Educational Multimedia, Hypermedia and Telecomunications,* 493–498 Chesapeake VA: AACE, Retrieved November 30, 2007, from http://infosoc.haifa.ac.il/DigitalLiteracyEshet.doc

Eshet-Alkalai, Y. (2004). Digital literacy: a conceptual framework for survival skills in the digital era. *Journal of Educational Multimedia and Hypermedia,* 1391, 93–106. Retrieved November 30, 2007, from http://www.openu.ac.il/Personal_sites/download/Digital-literacy2004-JEMH.pdf

Gardner, H. (2006). *Five Minds for the Future,* Boston, MA: Harvard Business School Press.

Gilster, P. (Ed.). (2006). Digital fusion: defining the intersection of content and communications. In A. Martin & D. Madigan (Eds.), *Digital literacies for learning* (pp. 42–50). London: Facet Publishing.

———. (1997). *Digital literacy.* New York: Wiley.

Hargittai, E. (2005). Survey measures of web-oriented digital literacy. *Social Science Computer Review, 23(3),* 371–379.

Inoue, H., Naito, E. & Koshizuka, M. (1997). Mediacy: what is it? *International Information and Library Review, 29(3/4),* 403–413.

Kauhanen-Simanainen, A. (2007). *Corporate literacy: discovering the senses of the organization.* Oxford: Chandos Publishing.

Kope, M. (Ed.). (2006). Understanding e-literacy. In A. Martin & D. Madigan (Eds.), *Digital literacies for learning* (pp. 68–79). London: Facet Publishing.

Lanham, R.A. (1995). Digital literacy, *Scientific American, 273(3),* 160–161.

Markless, S., & Streatfield, D. (2007). Three decades of information literacy: redefining the parameters. In S. Andretta (Ed.), *Change and challenge: information literacy for the 21st century* (pp.15–36). Adelaide: Auslib Press.

Martin, A. (2006a). Literacies for the digital age. In A. Martin & D. Madigan (Eds.), *Digital literacies for learning* (pp. 3–25). London: Facet Publishing.

———. (2006b). A framework for digital literacy, DigEuLit working paper. Retrieved November 15, 2007, from http://www.digeulit.ec/docs/public.asp

———. (2005). The landscape of digital literacy, DigEuLit working paper. Retrieved November 30, 2007, from http://www.digeulit.ec/docs/public.asp

———. (2003). Towards e-literacy. In A. Martin & H. Rader (Eds.), Information and IT literacy: enabling learning in the 21st century (pp. 3–23). London: Facet Publishing.

Martin, A., & Madigan, D. (Eds.). (2006a). *Digital literacies for learning*. London: Facet Publishing.

———. (Eds.). (2006b). Preface. In A. Martin & D. Madigan (Eds.), *Digital literacies for learning* (pp. xxv–xxviii). London: Facet Publishing.

McClure, C.R. (1994). Network literacy: a role for libraries. *Information Technology and Libraries*, *13(2)*, 115–125.

Neelameghan, A. (1995). Literacy, numeracy . . . informacy, *Information Studies*, 1(4), 239–249.

Nicholas, D., & Williams, P. (1998). Review of P. Gilster "Digital Literacy', *Journal of Documentation*, 54(3), 360–362.

Pilerot, O. (2006). Information literacy: An overview. In A. Martin & D. Madigan (Eds.), *Digital literacies for learning* (pp. 80–88). London: Facet Publishing.

Robinson, L., & Bawden, D. (2001). Libraries and open society: Popper, Soros and digital information. *Aslib Proceedings*, *53(5)*, 167–178.

Robinson, L., Kupryte, R., Burnett, P., & Bawden, D. (2000). Libraries and the Internet: overview of a multinational training course. *Program*, *34(2)*, 187–194.

Robinson, L. et al. (2005). Healthcare librarians and learner support: competences and methods. *Health Information and Libraries Journal*, 22 (supplement 2), 42–50.

SCONUL (2006). SCONUL Seven Pillars model for information literacy, Retrieved November 30, 2007, from: http://www.sconul.ac.uk/groups/information_literacy/seven_pillars.html

Shapiro, J.J. & Hughes, S.K. (1996). Information technology as a liberal art, *Educom Review*, *31(2)*, March/April 1996.

Snavely, L., & Cooper, N. (1997). The information literacy debate. *Journal of Academic Librarianship*, *23(1)*, 9–20.

Virkus, S. (2003), Information literacy in Europe: A literature review, *Information Research*, 8(4), paper 159, Retrieved November 30, 2007, from http://informationr.net/ir/8–4/paper159.html

Webber, S., & Johnson, B. (2000). Conceptions of information literacy: new perspectives and implications. *Journal of Information Science*, *26(6)*, 381–397.

Willams, P., & Minnian, A. (2007). Exploring the challenges of developing digital literacy in the context of special educational needs communities. In S. Andretta (Ed.), *Change and challenge: information literacy for the 21ˢᵗ century* (pp. 115–144). Adelaide: Auslib Press.

CHAPTER TWO

Functional Internet Literacy

Required Cognitive Skills with Implications for Instruction

GENEVIEVE MARIE JOHNSON

Introduction

The ability to sign one's name was once the benchmark of literacy. Over time the concept evolved to refer to functional reading and writing competencies (Tyner, 1998). Currently, literacy includes the ability to use a variety of technologies (Selber, 2004), although precise definitions are lacking. For example, *computer literacy* has been referred to as skill with spreadsheets and word processing (Reed, Doty, & May, 2005), programming and software applications (Wilson, 2003), internet and software applications (Hackbarth, 2002), internet, database, and graphics applications (Nokelainen, Tirri, & Campbell, 2002), and knowledge of security and hardware (Schaumburg, 2001). Eisenberg and Johnson (2002) suggested that the result of computer literacy is "to use technology as a tool for organization, communication, research, and problem solving" (p. 1). Gurbuz, Yildirim, and Ozden (2001) argued that "the definition of computer literacy will be specific to the context in which the computer-literate person must function" (p. 260). Talja (2005) noted that "definitions of computer literacy are often mutually contradictory". (p. 13)

Contemporary conceptualization of literacy also involves technology, information, and communication literacy. *Technology literacy* is defined as "an individual's abilities to adopt, adapt, invent, and evaluate technology to positively affect his or her life, community, and environment" (Hansen, 2005, p. 1). Weber (2005) claimed that technology literacy education is based on the assumption that "all persons must be knowledgeable of their technological environment so they can participate in controlling their own destiny" (p. 29). McCade (2001) defined "information literacy" as the capacity to access and evaluate "information from a variety of both electronic and non-electronic sources" (p. 1). The Educational Testing Service (2005) uses the term "information and communication literacy" to refer to "abilities to find, use, manage, evaluate and convey information efficiently and effectively" (p. 1).

While implicit in current views of literacy (Selber, 2004), "internet literacy" remains an ambiguous term. "Internet information literacy" has been defined as the capability to access and evaluate online information (Eisenberg & Johnson, 2002; O'Sullivan & Scott, 2000a, 2000b). Harkham Semas (2002) used the term "internet literacy" to specifically refer to online search competence. Hofstetter (2003) provided a more comprehensive perspective on internet literacy as including skill with connectivity, security, communication, multimedia, and web page development. While web page development is a useful skill, it is not a *functional* skill; that is, it is not a requirement of typical internet use (Australian Government, 2005; Statistics Canada, 2004; U.S. Census Bureau, 2005).

"Functional literacy" reflects typical use and common requirements (Selber, 2004). Thus, "functional internet literacy" is defined in terms of online activities common to the majority of users. In a democracy, definition of functional literacy is prerequisite to distinguishing the literate from the illiterate, evaluating the consequences of such distinction, and rectifying any identified disadvantage (Papen, 2005).

Patterns of Internet Use and Functional Internet Literacy

Nie, Simpser, Stepanikova, and Zheng (2005) reported that 57% of internet use relates to communication (i.e., email, instant messaging, and chat) and 43% involves browsing (i.e., visiting web sites including those with message boards). Such dichotomization of online activities is curious since message boards are typically viewed as asynchronous communication tools (Branon & Essex, 2001). Williams (2001) suggested that all internet activities (e.g., banking and shopping) are communicative because information is exchanged—the standard definition of communication (Alder, Rosenfeld, & Proctor, 2004). While there is disagreement regarding what constitutes internet communication, email is frequently cited as the most common online activity (Australian Government, 2005; Rotermann, 2001; Statistics Canada, 2004; UCLA World Internet Project, 2004; U.S. Census Bureau, 2005).

Following communication, accessing information is reportedly the second most common online activity (Nie et al., 2005). Approximately half of all internet users have searched for health-related information (Stevenson, 2002), and half of all adolescents have obtained online information specifically related to reproductive health (Borzekowski & Rickert, 2001). Nearly 80% of internet users go online to locate information on specific products and services; almost 70% access news, weather, and sports information (U.S. Census Bureau, 2005). Approximately one-third of the time that children are online, they report accessing web sites (Roberts, Foehr, & Rideout, 2004). Two-thirds of parents maintain that access to information related to schoolwork is the primary advantage of their children's internet use (Clark, 2001). Indeed, the internet is virtually synonymous with information location and retrieval (Karchmer, 2001; Tesdell, 2005).

The internet is also a common source of recreation (Nie et al., 2005). Approximately one-third of the time that children are online, they report playing games (Roberts et al., 2004). The majority of North American adolescents (Rotermann, 2001) and half of China's 100 million internet users play online games (Martinsons, 2005). More than one-third of adult internet users play online games and 21% watch online movies and listen to online music (U.S. Census Bureau, 2005). Surprisingly, 42% of heavy online gamers are over the age of 35 (Hopper, 2002). According to Silver (2001), 80% of senior citizens who access the internet "go online for personal interest and entertainment" (p. 9). Jones Thompson (2005) reported that sites associated with music, games, movies, videos, pornography, and gambling or sweepstakes are characterized by the heaviest internet traffic.

The internet is increasingly used by the general population for commercial purposes (Shop.org, 2005). More than half of internet users have made a purchase online (U.S. Census Bureau, 2005). In 2003, approximately 57% of households using the internet had someone who banked online, a significant increase from recent years. "This growth may indicate consumers are becoming more confident in the internet's security aspects" (Statistics Canada, 2004,

p. 2). Jones Thompson (2005) reported that most internet spending is associated with dating, entertainment, investments, research, personal growth, and games. Ellison and Clark (2001), however, identified print material (e.g., books and magazines), computer software, music (e.g., CDs and MP3), and travel arrangements as the most frequently occurring e-commerce transactions. LaRose (2001) demonstrated that popular e-commerce sites often include features that stimulate unregulated buying and that impulsive shopping accounts for one-fourth of e-commerce purchases (LaRose & Eastin, 2002).

Figure 2.1. Categorization of Common Internet Activities.

Category	Internet Activities
Communication	• bulletin boards • chat and email • instant message
Information	• health • personal • interest
Recreation	• movies • music • games
Commercial	• bank and invest • purchase products • access services
Technical	• connectivity • security • downloads

Categorization of common internet activities may be arbitrary and category overlap is apparent (Nie et al., 2005; Williams, 2001). Nonetheless, patterns of daily use support the organization of internet activities in terms of communication, information, recreation, and commercial pursuits. Additionally, technical competence with connectivity, security, and downloads is prerequisite to internet use (Hofstetter, 2003). Figure 2.1 provides a graphic representation of categories of common online activities and the related requirement of technical skill. Such categorization reflects common internet use in daily life and provides a mechanism for organizing activities for which individuals require a level of functional literacy. Functional use of all categories of common internet activities requires a range of cognitive skills.

Functional Internet Literacy: Required Cognitive Skills

Bloom's taxonomy of cognitive skills has guided educators for 50 years (Bloom, Engelhart, Frost, Hill, & Krathwohl, 1956) and is considered one of the most significant educational contributions of the 20th century (Anderson & Sosniak, 1994). Bloom's taxonomy (1984) includes: *knowledge* (i.e., remembering and recognizing), *comprehension* (i.e., understanding), *application* (i.e., using a general concept to solve a specific problem), *analysis* (i.e., understanding the components of a larger process or concept), *synthesis* (i.e., combining ideas and information), and *evaluation* (i.e., judging value or quality). The taxonomy reflects a "complexity hierarchy that orders cognitive processes from simply remembering to higher order critical and creative thinking" (Noble, 2004, p. 194). Figure 2.2 provides a summary of the taxonomy including operational definitions for each of the six levels of cognitive processing. Such a hierarchy provides a structure for organizing the cognitive skills required by users who are functionally internet literate.

As illustrated in Figure 2.3, each category of functional internet use demands a range of cognitive skills. For example, when users engage in online communication in the form of email, basic *knowledge* (i.e., email addressing) is necessary. Further, the online communicator must read a message with *comprehension*. A deep level of message comprehension is demonstrated when the receiver acts upon the message, for example, complies with a request (i.e., *application*). To effectively comply with an email request, the receiver discriminates between essential and nonessential aspects of the message (i.e., *analysis*). *Synthesis* is required to collectively interpret multiple email messages. Finally, the highest level of cognitive processing requires *evaluation* (e.g., email sender intention). In the context of email communication, physical gestures, facial expressions, and spontaneity are largely absent; evaluating digital messages may be more intellectually challenging than evaluating face-to-face communication (Szewczak & Snodgrass, 2002).

Figure 2.2. Taxonomy of Cognitive Skills (adapted from Bloom, 1984).

Cognitive Skill	Specific Demonstration
Knowledge	• recall of specific fact • recognize associations • list, define, identify, quote, indicate • show, label, collect, name, tabulate
Comprehension	• understand information • interpret, compare, contrast • order, group, summarize, describe • distinguish, estimate, extend, discuss
Application	• use information in daily life • generalize to new situations • solve problems and demonstrate • complete, illustrate, discover
Analysis	• identify patterns and relationships • separate and organize components • infer meanings and explain • order, connect, arrange, divide
Synthesis	• generalize and integrate information • predict, conclude, substitute, plan • combine, modify, rearrange, create • design, invent, compose, prepare
Evaluation	• verify information and decide • assess evidence and conclude • make choices based on reason • rank, recommend, judge, support

Low-level knowledge skills without corresponding higher-order thinking skills render the user ill-equipped to analyze, synthesize, and evaluate common online activities (Kay & Delvecchio, 2002; Rogers & Swan, 2004; Younger, 2005). To be knowledgeable about search engines but unable to discriminate between fact and opinion renders the user susceptible to misinformation;

the internet has been described as a repository of opinion disguised as fact (O'Sullivan & Scott, 2000a, 2000b). Correspondingly, to be knowledgeable about chat rooms but unable to discriminate friend from foe renders the user vulnerable to exploitation; the internet has been described as a hunting ground for human predators (Finkelhor, Mitchell, & Wolak, 2000; Katz & Rice, 2002). In this regard, functional internet literacy cannot be conceptualized simply in terms of basic knowledge skills. Such a definition places internet users at risk for a variety of adverse online experiences.

Figure 2.3. Required Cognitive Skills for Common Internet Use.

Internet Use	Cognitive Skill	Demonstration of Cognitive Skill
Communication	Knowledge Comprehension Application Analysis Synthesis Evaluation	• enter email address • read and understand message • comply with email instructions • identify essential information • combine multiple messages • judge intention of communicator
Information	Knowledge Comprehension Application Analysis Synthesis Evaluation	• enter search term • understand site navigation • use information in daily life • determine site author • summarize site information • evaluate site information
Recreation	Knowledge Comprehension Application Analysis Synthesis Evaluation	• enter game site URL • understand game rules • locate similar games online • determine game elements • generalize skill from other game • judge quality of game

(*Figure 2.3. continued on following page*)

Internet Use	Cognitive Skill	Demonstration of Cognitive Skill
Commercial	Knowledge Comprehension Application Analysis Synthesis Evaluation	• enter credit card information • understand product information • compare with real products • identify critical components • combine multi-site information • decide on the best product
Technical	Knowledge Comprehension Application Analysis Synthesis Evaluation	• launch browser • explain importance of browser • use alternative browser • compare browser features • align browser with requirements • recommend browser

Defining functional internet literacy in terms of a hierarchy of cognitive skills accounts for age and education differences in patterns of use and degree of online vulnerability. Young users often have basic knowledge skills which support internet access (Clark, 2001) but lack the cognitive maturity to evaluate online communication and information (Katz & Rice, 2002). Many senior citizens reportedly lack basic knowledge of internet operation (Lundt & Vanderpan, 2000; Silver, 2001) but, given access, have the higher-order thinking skills prerequisite to effective online interaction and transaction (Reed et al., 2005). Correspondingly, level of formal education is consistently associated with internet use; as amount of education increases, internet use increases (Australian Government, 2005; Statistics Canada, 2004; U.S. Census Bureau, 2005). Maturation (i.e., age) and experience (e.g., education) contribute to the development of higher-order thinking skills (Yan, 2005), which, in turn, contribute to the development of functional internet literacy.

The Development of Functional Internet Literacy

Based on direct observation of children in India exposed to computers for the first time, Mitra and Rana (2001) concluded that computing and internet skills emerge without formal instruction. Dryburgh (2002) reported that the vast majority of computer users learn necessary skills informally (e.g., trial-and-error, help from a friend) and semi-formally (e.g., online tutorial, manual). Silver, Williams, and McOrmond (2001) reported that in the 1990s the most popular topic of independent study was computer and internet technologies. Harkham Semas (2002) observed that children's "educational use of the internet mainly occurs outside the school day with little direction, if any, from teachers" (p. 11). Based on extensive interviews with children, Burnett and Wilkinson (2005) concluded that "school internet use was geared towards practice for the real world, whilst home use was embedded in life" (p. 159).

Hackbarth (2002) maintained that the "impact of relying upon home rather than classroom to enhance computer literacy" has resulted in "inequitable access" and "unequal achievement" (p. 53). Many skills, particularly higher-order thinking skills, are best achieved in formal learning environments (Karasavvidis, Pieters, & Plomp, 2003). Thus, functional internet literacy, which requires complex cognitive processing, is best achieved in structured and directed learning situations. Formal teaching of functional internet literacy is based upon assessment of cognitive skills deficits and instruction that targets those identified deficits.

Functional Internet Literacy: Assessment and Targeted Instruction

Reporting that university students often lack confidence in their internet competencies, Messineo and DeOllos (2005) encouraged instructors to "better target their efforts by conducting skill surveys early in the term" (p. 53). As well, recently available commercial tests assess a range of cognitive skills prerequisite to effective internet use. For example, the *Information and Communication Technology Literacy Assessment* is a simulation-based test that measures user ability to access, manage, integrate, evaluate, create, and communicate information in technical contexts (Murray, 2005). The test measures "not just knowledge of technology, but the ability to use critical-thinking skills to solve problems within a technological environment" (Educational Testing Service, 2005, p. 1). Bloom's taxonomy (1984) provides a structure for organizing the cognitive and critical-thinking skills necessary for functional internet literacy.

Prior to instruction, an inventory of required internet skills is developed. Such an inventory may be based on categorization of common internet use presented in Figure 2.1 or identification of required internet activities in a specific context or for a specific individual. Required skills may be determined via analysis of user activity, observation of users, or interviews with managers. The inventory results in a checklist that includes all levels of cognitive skills (i.e., knowledge, comprehension, application, analysis, synthesis, and evaluation) for all required online activities. The checklist, completed by a manager, director, teacher or user, facilitates identification of cognitive skill deficits (i.e., skills required by the user that are not adequately developed). In some cases, higher-order cognitive skills may be difficult to assess if low-level skill deficits restrict internet access. The checklist forms the basis for *targeted instruction*.

Targeted instruction refers to teaching efforts directed specifically toward identified skill deficits (Dreher, 2001). In the case of functional internet literacy, instruction is sequenced to address the identified skill deficits lowest on the cognitive hierarchy; low-level skills are often prerequisite to higher levels of cognitive processing (Bloom, 1984). For example, the checklist of internet requirements developed for a specific context (e.g., a retirement community) is administered to relevant individuals (e.g., those interested in using technology). Individuals with identified knowledge skill deficits are grouped for instruction. Demonstration and practice with basic internet knowledge skills continues until a criterion is met (e.g., independent use of email). Subsequently, individuals with identified comprehension skill deficits are grouped for instruction. Discussion and demonstration of internet comprehension skills continue until a criterion is met (e.g., correctly paraphrase email messages). Targeted instruction progresses up the hierarchy of cognitive skills for all online activities commonly required by the user.

Patterns of daily internet use form the basis of functional internet literacy. Bloom's taxonomy of cognitive skills (1984) provides a structure by which to organize the intellectual requirements of effective online communication, information, recreation, and commercial pursuits and the technical skills necessary to operate the equipment that mediates such activities. A comprehensive hierarchy of cognitive skills applied to common online activities directs assessment and targets instruction. Functional internet literacy is not the ability to use a set of technical tools; rather, it is the ability to use a set of cognitive tools.

References

Alder, R. B., Rosenfeld, L. B., & Proctor, R. F. (2004). *Interplay: The process of interpersonal communication*. New York: Oxford University Press.

Anderson, L. W., & Sosniak, L. A. (1994). *Bloom's taxonomy: A forty-year retrospective. Ninety-third yearbook for the National Society for the Study of Education, Part II*. Chicago: University of Chicago Press.

Australian Government. (2005). *The current state of play*. Canberra, Australia: Department of Communications, Information Technology, and the Arts. Retrieved November 30, from http://www.dcita.gov.au/__data/assets/pdf_file/33120/CSP2005.pdf

Bloom, B. S. (1984). *Taxonomy of educational objectives*. Boston: Pearson Education.

Bloom, B. S., Engelhart, M. D., Frost, E. J., Hill, W. H., & Krathwohl, D. R. (1956). *Taxonomy of educational objectives: Handbook 1, cognitive domain*. New York: David McKay.

Borzekowski, D. L., & Rickert, V. I. (2001). Adolescents, the Internet, and health: Issues of access and content. *Journal of Applied Developmental Psychology, 22*, 49–59.

Branon, R. F., & Essex, C. (2001). Synchronous and asynchronous communication tools in distance education: A survey of instructors. *TechTrends, 45*, 36, 42.

Burnett, C., & Wilkinson, J. (2005). Holy lemons! Learning from children's uses of the Internet in out-of-school contexts. *Literacy, 39*, 158–164.

Clark, W. (2001). Kids and teens on the Net. *Canadian Social Trends, 62*, 6–10.

Dreher, M. J. (2001). Children searching and using information text: A critical part of comprehension. In C. Collins Block & M. Pressley (Eds.), *Comprehension instruction: Research-based best practices* (pp. 289–305). New York: Guilford.

Dryburgh, H. (2002). Learning computer skills. *Canadian Social Trends, 64*, 20–24.

Educational Testing Service. (2005). *ICT Literacy Assessment—Information and Communication Technology Literacy*. Retrieved November 30, 2007, from http://www.ets.org/ictliteracy.

Eisenberg, M. B., & Johnson, D. (2002). Learning and teaching information technology–Computer skills in context. *ERIC Digest*. (ERIC Document Reproduction Service No. ED 465 377).

Ellison, J., & Clark, W. (2001). Net shopping. *Canadian Social Trends, 60*, 6–9.

Finkelhor, D., Mitchell, K. J., & Wolak, J. (2000). *Online victimization: A report on the nation's youth*. Alexandria, VA: National Center for Missing and Exploited Children.

Gurbuz, T., Yildirim, S., & Ozden, M. Y. (2001). Comparison of on-line and traditional computer literacy courses for preservice teachers: A case study. *Journal of Educational Technology Systems, 29*, 259–269.

Hackbarth, S. L. (2002). Changes in 4th graders' computer literacy as a function of access, gender, and email networks. *TechTrends, 46*, 46–55.

Hansen, J. (2005). *How would you know if you saw a technologically literate person?* Center for Technology Literacy. Retrieved December 4, 2007, from http://www.texastechnology.com/CTL_Research/5_Pillars

Harkham Semas, J. (2002). Digital disconnect: Teens say teachers lack Internet skills. *District Administration, 38*, 11.

Hofstetter, F. (2003). *Internet literacy* (3rd ed.). Boston: Irwin-McGraw-Hill.

Hopper, J. (2002). The (online) games people play. *Brandweek, 43*, 16.

Jones Thompson, M. (2005). Online recreation. *Technology Review, 108*, 32.

Karasavvidis, I., Pieters, J. M., & Plomp, T. (2003). Exploring the mechanisms through which computers contribute to learning. *Journal of Computer Assisted Learning, 19*, 115–128.

Karchmer, R. A. (2001). The journey ahead: Thirteen teachers report how the Internet influences literacy and literacy instruction in their K—12 classrooms. *Reading Research Quarterly, 36*, 442–466.

Katz, J. E., & Rice, R. E. (2002). *Social consequences of internet use: Access, involvement, and interaction.* Cambridge, MA: MIT Press.

Kay, H., & Delvecchio, K. (2002). *The world at your fingertips: Learning & internet skills.* Fort Atkinson, WI: Upstart.

LaRose, R. (2001). On the negative effects of e-commerce: A sociocognitive exploration of unregulated on-line buying. *Journal of Computer Mediated Communication, 16.* Retrieved November 30, 2007, from http://jcmc.indiana.edu/vol10/issue1/kim_larose.html

LaRose, R., & Eastin, M. S. (2002). Is online buying out of control? Electronic commerce and consumer self-regulation. *Journal of Broadcasting and Electronic Media, 46*, 549–565.

Lundt, J. C., & Vanderpan, T. (2000). It computes when young adolescents teach senior citizens. *Middle School Journal, 31*, 18–22.

Martinsons, M. G. (2005). Online games transform leisure time for young Chinese. *Communications of the ACM, 48*, 51.

McCade, J. M. (2001). Technology education and computer literacy. *The Technology Teacher, 61*, 9–13.

Messineo, M., & DeOllos, I. Y. (2005). Are we assuming too much? Exploring students' perceptions of their computer competence. *College Teaching, 53*, 50–55.

Mitra, S., & Rana, V. (2001). Children and the Internet: Experiments with minimally invasive education in India. *British Journal of Educational Technology, 32*, 221–232.

Murray, J. (2005). Testing information literacy skills (grades K—12). *Information Literacy for the Information Age.* Retrieved November 30, 2007, from http://www.big6.com/showarticle.php?id=484

Nie, N. H., Simpser, A., Stepanikova, I., & Zheng, L. (2005). *Ten years after the birth of the Internet, how do Americans use the Internet in their daily lives?* Stanford Center for the Quantitative Study of Society. Retrieved November 30, 2007, from http://www.stanford.edu/group/siqss

Noble, T. (2004). Integrating the Revised Bloom's Taxonomy with Multiple Intelligences: A planning tool for curriculum differentiation, *Teachers College Record, 106*, 193–211.

Nokelainen, P., Tirri, K., & Campbell, J. R. (2002, April). *Cross-cultural findings of computer literacy among the academic Olympians.* Paper presented at the Annual Meeting of the American Educational Research Association, New Orleans, LA. (ERIC Document Reproduction Service No. ED 467 500).

O'Sullivan, M., & Scott, T. (2000a). Teaching Internet information literacy: A critical evaluation. *Multimedia Schools, 7*, 40–42, 44.

———. (2000b). Teaching Internet information literacy: A collaborative approach. *Multimedia Schools, 7*, 34–37.

Papen, U. (2005). *Adult literacy as social practice: More than skills.* New York: Routledge.

Reed, K., Doty, D. H., & May, D. R. (2005). The impact of aging on self-efficacy and computer skill acquisition. *Journal of Managerial Issues, 17*, 212–229.

Roberts, D. F., Foehr, U. G., & Rideout, V. (2004). *Generation M: Media in the lives of 8–18 year olds*. Menlo Park, CA: The Henry J. Kaiser Family Foundation. Retrieved November 30, 2007, from http://www.kff.org/entmedia/loader.cfm?url=/commonspot/security/getfile.cfm&PageI=51809

Rogers, D., & Swan, K. (2004). Self-regulated learning and Internet searching. *Teacher's College Record, 106*, 1804–1824.

Rotermann (2001). Wired young Canadians. *Canadian Social Trends, 63*, 4–8.

Schaumburg, H. (2001, July). *Fostering girls' computer literacy through laptop learning: Can mobile computers help to level out the gender difference?* Paper presented at the National Educational Computing Conference, Chicago, IL (ERIC Document Reproduction Service No. ED 462 945).

Selber, S. A. (2004). *Multiliteracies for a digital age*. Carbondale, IL: Southern Illinois University Press.

Shop.org. (2005). *Statistics: US Online Shoppers*. The Network for Retailers Online. Retrieved November 30, 2007 from http://www.shop.org/learn/stats_usshop_general.asp.

Silver, C. (2001). Older surfers. *Canadian Social Trends, 63*, 9–12.

Silver, C., Williams, C., & McOrmond, T. (2001). Learning on your own. *Canadian Social Trends, 60*, 19–21.

Statistics Canada. (2004). *Household Internet use survey*. Retrieved November 30, 2007 from http://www.statcan.ca/Daily/English/040708/d040708a.htm

Stevenson, K. (2002). Health information on the Net. *Canadian Social Trends, 66*, 7–10.

Szewczak, E. J., & Snodgrass, C. R. (2002). *Managing the human side of information technology: challenges and solutions*. Hershey, PA: The Idea Group.

Talja, S. (2005). The social and discursive construction of computing skills. *Journal of the American Society for Information Science and Technology, 56*, 13–22.

Tesdell, L. S. (2005). Evaluating web-sources: Internet literacy and L2 academic writing. *Technical Communication, 52*, 401.

Tyner, K. (1998). *Literacy in a digital world—Teaching and learning in the age of information*. Mahwah, NJ: Lawrence Erlbaum.

UCLA World Internet Project. (2004). First release of findings from the UCLA World Internet Project shows significant 'Digital Gender Gap' in many countries. *UCLA News*. Retrieved December 2, 2007, from http://newsroom.ucla.edu/page.asp?RelNum=4849

U.S. Census Bureau. (2005). *Computer and internet use in the United States: 2003*. Retrieved November 27, 2005, from http://www.census.gov/prod/2005pubs/p23–208.pdf

Weber, K. (2005). A proactive approach to technology literacy. *The Technology Teacher, 64*, 28–30.

Williams, C. (2001). The evolution of communication. *Canadian Social Trends, 60*, 15–18.

Wilson, F. (2003). Can compute, won't compute: Women's participation in the culture of computing. *New Technology, Work, and Employment, 18*, 127–142.

Yan, Z. (2005). Age differences in children's understanding of the complexity of the Internet. *Applied Developmental Psychology, 26*, 385–396.

Younger, P. (2005). The effective use of search engines on the internet. *Nursing Standards, 19*, 56–64.

CHAPTER THREE

Digital Literacy as Information Savvy

The Road to Information Literacy

MAGGIE FIELDHOUSE AND DAVID NICHOLAS

Introduction

The digital revolution has changed information-seeking behavior beyond recognition and, by means of simple to use web interfaces and search engines, has rendered us all information "savvy," in that we think we can easily find, create and use information on the internet. In this chapter, we explore what it means to be information savvy and define digital and information literacies, examining the relationships between the underlying technological skills of digital literacy, which we use to interact with internet tools such as search engines and those of information literacy, which enable us to evaluate and make relevant judgments about what we find.

The ability to interact with computers to locate information has been the focus of much research in the last twenty-five years, and this considerable body of work is reviewed to trace the influence of technology on information seeking behavior. Research carried out at the Centre for Interactive Behaviour and the Evaluation of Research (CIBER) at University College, London shows that typical information-seeking behavior tends to be simplistic, based on the construction of simple searches and characterized by a tendency to "bounce" from one web page to another to find information. Such information-seeking behavior patterns raise questions about whether being information savvy is enough in an increasingly information-dependent society, and what impact the principles of information literacy might have on how we locate and use information to create new knowledge.

This chapter explains how those principles, which encourage us to think more critically about the information that we find, are seen by information professionals as a means of counteracting what might be seen as a "dumbing down" of information-seeking behavior as a direct result of its migration to the virtual environment. The internet presents us with a bewildering array of information choices, and we devise coping strategies according to our level of information savviness and our digital literacy skills to avoid being overwhelmed.

What Does Being Information Savvy Mean?

So what is being information savvy all about? The *Oxford English Dictionary* defines "savvy" as "having practical sense, quick-witted; knowledgeable, wily, experienced. Also wise to (something)." In a digital environment, being information savvy is more than just being able to use technology to locate information. It suggests a common sense approach to and awareness of the problems and pitfalls of exploring the highways of the internet, just as being street-wise implies being able to handle the harsher realities of city life.

But is savviness enough, or do we need more than that to successfully navigate the virtual environment and find appropriate information? Such navigation skills become increasingly significant as the need to become informed citizens through lifelong learning pervades education, the workplace, and our personal lives. The importance of making successful information choices perhaps renders being "savvy" insufficient, and we need to develop our knowledge and experience further so that we can make value judgments about the quality and relevance of information and become information "wise." The *Oxford English Dictionary* defines "wise" as

> [h]aving or exercising sound judgement or discernment; capable of judging truly concerning what is right or fitting, and disposed to act accordingly; having the ability to perceive and adopt the best means for accomplishing an end; characterised by good sense and prudence.

Being practical, quick-witted and wily captures the essence of being information savvy, but the ability to exercise judgment, discernment, and prudence reflects the more literate and thoughtful approach to information seeking and handling that enables us to become information wise.

A whole generation in developed countries has grown up in a digital society, exposed to vast amounts of information in a variety of formats: text, images, video and audio. In an electronic world these digital natives could be considered to be "information savvy" as well as digitally literate. They interact naturally with technologies such as instant messaging, blogs, wikis, video games, social networking tools such as MySpace, YouTube and Facebook, commercial websites such as Amazon and iTunes and of course, Google, the popular search engine of choice. They combine work and social life, instinctively searching the internet and freely navigating through unstructured, non-sequential links to locate information of all kinds while simultaneously emailing, chatting with peers, playing games and working on assignments. Such freedom evokes images of the internet as a giant sweetshop, in which we behave like children, grabbing all we can get with less regard for quality than quantity.

We all know, though, that this sort of chaotic behavior has, ideally, to be checked if we are to become responsible and discriminating members of society. So we learn how to be selective, which, in information terms, means considering the quality, value and reliability of what we find by applying higher order skills such as critical thinking. These higher order skills have historically been taught by library and information workers who have trained past generations of users in how to find and use information. Dating back to the days of tuition in using library card catalogues, librarians have guided students and the public through the maze of published and unpublished material and have acted as intermediaries in the information-seeking process. In an environment dominated by search engines and easy access to vast quantities of information, do we still need intermediaries to help us become "information wise" through information literacy programs designed to develop the evaluative skills that enable us to distinguish between good and bad information? Or is information savviness enough?

Digital and Information Literacy Defined

Definitions of digital and information literacy are numerous. Within this pool of definitions, terms often are interchangeable; for example, "literacy," "fluency" and "competency" can all be used to describe the ability to steer a path through digital and information environments to find, evaluate, and accept or reject information. In what follows, we provide an overview of—and discuss—various definitions of digital literacy and information literacy currently available.

Digital Literacy

While much is written in the name of digital literacy, consensus on a single definition of the phrase seems to be elusive. Martin and Madigan (2006), for example, explore a range of conceptions of digital literacy and how these conceptions are enabled and supported in different communities. A broad definition of "digital literacy" nonetheless is provided by Martin (2005), who acknowledges related "literacies," such as ICT literacy, information literacy, media literacy and visual literacy which have gained new or increased relevance in the digital environment. He describes digital literacy as "the ability to succeed in encounters with the electronic infrastructures and tools that make possible the world of the twenty-first century." (2005, p. 131) Concerned with digital literacy and e-learning, Martin sees the need for mastering electronic tools as crucial to success in learning communities. He also contends that digital literacy involves "acquiring and using knowledge, techniques, attitudes and personal qualities and will include the ability to plan, execute and evaluate digital actions in the solution of life tasks, and the ability to reflect on one's own digital literacy development." (2005, p. 135)

This view suggests that digital literacy is a prerequisite for learning in a student centered educational culture, but the question remains: is this enough? As well as being digitally literate, we argue that students need to be "information savvy" and capable of identifying when information is needed, how to locate it, and how to use it effectively. These abilities are an equally important, fundamental part of the learning process in formal and informal education as well as essential for lifelong learning. Commenting on digital literacy as a key area of competence in schools, digital literacy for school-age learners in Norway is defined by Ola Erstad (2006, p. 416) as ". . . . .skills, knowledge and attitudes in using digital media to be able to master the challenges in the learning society." However, the role of schools in the development of digital literacy is an interesting, if inconclusive one, and factors such as policies and rules, technological and filtering controls and time constraints affect teachers' abilities to support students in maximizing the potential of the internet in educational activities. Students also bring very mixed abilities, attitudes to learning, and personal attributes to the classroom, and experiences in primary school will affect the development of literacy skills (Fidel et al.,1999; Selwyn, 2006; Madden et al., 2006; Pew Internet and American Life Project 2002a, 2002b; Williams & Wavell, 2006).

A rather different view of digital literacy is presented by Paul Gilster (1997), who writes about the digital revolution in his book, *Digital Literacy*. Gilster introduced the idea that being digitally literate pertains to the cognitive process of using electronic information, defining it as "the ability to understand and use information in multiple formats from a wide range of sources when it is presented via computers." (p. 33) Throughout *Digital Literacy*, he celebrates the benefits of developments in e-information, charting such benefits across day-to-day online activities such as emailing, catching up with newsgroup postings, checking investments, making travel arrangements, and keeping up to date with news stories. Although Gilster's position has been criticized for presenting the subject in a naïvely optimistic and somewhat trivial manner (e.g., Nicholas & Williams, 1998), Bawden (2001, p. 249) acknowledges that while already well known within the confines of the library and information community, Gilster's ideas on digital literacy have also "had a considerable impact on the wider sphere."

Gilster's book raises two important points about digital literacy which are central to the concept of being information savvy:

- he describes how the digital environment has revolutionized not only information-seeking, but also information-handling behavior.
- he suggests that technical skills may be less important than a discriminating view of what is found on the internet.

While not making any attempt to provide structured lists of specific skills for digital literacy, Gilster does identify four core competencies:

- knowledge assembly
- internet searching
- hypertextual navigation
- content evaluation.

As we shall see, these competencies are reflected in those defined for information literacy and are illustrated in Table 3.1.

Information Literacy

Definitions of information literacy are more substantial and have been adopted at a national level in the U.S., Australia, New Zealand, and in the U.K. In 1989, the Final Report of the American Library Association Presidential Committee on Information Literacy stated that,

> To be information literate, a person must be able to recognize when information is needed and have the ability to locate, evaluate, and use effectively the needed information.

This early definition was extended by the Association of College and Research Libraries (ACRL) in 2001 to become the Competency Standards for Higher Education, which provides a framework of five primary standards, 22 performance indicators, and 87 outcomes for assessing the information-literate individual.

In the U.K., the Society of College, National and University Libraries (SCONUL) defined the Seven Pillars Model for Information Literacy in 1999 and, in 2004, CILIP, the Chartered Institute of Library and Information Professionals, agreed upon a definition of information literacy as

> knowing when and why you need information, where to find it, and how to evaluate, use and communicate it in an ethical manner.

The Council of Australian University Librarians' (CAUL) Information Literacy Standards (2001) and the later Australia and New Zealand Institute for Information Literacy's framework (see Bundy, 2004) extended the American Library Association's definition to include an understanding of the "economic, legal, social and cultural issues in the use of information" and the recognition of "information literacy as a prerequisite for lifelong learning."

Gilster's concept of "competence with knowledge assembly" represents the bringing together of existing and new knowledge and can be considered as functioning at two levels in the realm of information literacy:

- the *pre-search* activity of gathering what is already known in order to identify knowledge gaps
- the *post-search* activity of organizing, managing and processing the newly-found information in such a way as to build new knowledge.

Table 3.1. Summary of Relationships Between Different Positions on Digital and Information Literacy.

Gilster's Digital Literacy Competencies	SCONUL Seven Pillars Model of Information Literacy (1999) (UK)	ACRL Information Literacy Competency Standards for Higher Education (2000) (U.S.)	ANZIIL Framework for Information Literacy (2004) (Australia and New Zealand)
Knowledge assembly (pre search)	Recognize a need for information Distinguish ways in which the information 'gap' may be addressed	Determine the nature and extent of the information needed	Recognize the need for information Determine the nature and extent of the information needed
Knowledge assembly (post search)	Organize, apply and communicate information to others in ways appropriate to the situation Synthesize and build upon existing information, contributing to the creation of new knowledge	Use information effectively, individually or as a member of a group, to accomplish a specific purpose Understand the economic, legal and social issues surrounding the use of information and access and use information ethically and legally	Manage information collected or generated Apply prior and new information to construct new concepts or create new understandings Use information with understanding and acknowledge cultural, ethical, economic, legal, and social issues surrounding the use of information

(Table 3.1. Continued on following page)

Gilster's digital literacy competencies	SCONUL Seven Pillars Model of Information Literacy (1999) (UK)	ACRL Information Literacy Competency Standards for Higher Education (2000) (U.S.)	ANZIIL framework for Information Literacy (2004) (Australia and New Zealand)
Internet searching	Locate and access information	Access the needed information effectively and efficiently	Find needed information effectively and efficiently
Hypertextual navigation			
Content evaluation	Compare and evaluate information obtained from different sources	Evaluate information and its sources critically and incorporate selected information into his or her knowledge base and value system	Critically evaluate information and the information-seeking process

(*Table 3.1. Continued from previous page*)

In the information-seeking environment there is considerable overlap between Gilster's digital competencies of internet searching and hypertextual navigation. In the digital world, internet searching involves the use of tools such as search engines to locate information, while hypertextual navigation represents the process of exploring links in an unstructured, non-linear e-space. In terms of information literacy, these competencies are more closely related, and are embodied in the activities of locating and accessing information. Together they are important influences on information seeking and handling behaviors in a digital environment, since the use of search tools together with navigation between links forms a single, intrinsic part of the search process, rather than the distinctly separate activities identified by Gilster.

The concept of content evaluation is common to both digital and information literacy and of fundamental importance to the idea of transforming the deficiencies of information savviness into information wisdom through information literacy. In the digital universe, access to the internet means that information-seeking activities can take place anywhere at any time, in isolation or collaboratively, rather than in the more confined, but social context of public or academic libraries. Today's digitally literate students prefer to find information for themselves using search engines, peers or chat rooms rather than by seeking advice from teachers or librarians, and they can quickly and easily assemble disassociated segments of information or misinformation to create new knowledge. This suggests a need for the critical thinking skills that information literacy encourages, determining how credible information is and to contextualize, analyze, and synthesize what is found online. In addition, ethical and legal considerations in the form of respect for copyright and intellectual property are necessary to ensure that information is used appropriately.

Being digitally literate and information savvy are not, on their own, enough to be considered information literate, and the authority, relevance, currency, quality, coverage and objectivity of information have to be assessed according to the defined standards.

Developing Digital Literacy and Information Wisdom: Following the Literature Trail

An examination of the published literature from the last 25 years demonstrates how the growth in access to electronic information has increased digital literacy through familiarity with a progressively sophisticated range of technologies and tools as well as introducing opportunities for interacting with information as consumers as well as creators.

The literature trail reflects demographic changes and emphasizes the differences in generational approaches to information technologies, digital literacy and information-seeking behavior. While there is little consensus on the terminology or precise time periods by which each generation is defined, they can be broadly categorized by technological developments, life experiences and political and social attitudes.

In his book, *Growing up Digital*, which profiles the rise of the Net Generation, Don Tapscott (1998) identifies three key groups in illustrating the relationships between demographical and technical changes:

- the Boom generation born between 1946 and 1964 whose lives were influenced by the expansion of television,
- the Bust generation, born between 1965 and 1976 and a time of low birth rates and economic downturn
- the Baby Boom Echo, the generation born between 1977 and 1997 which has, as a result of the digital revolution, become known as the Net Generation.

Baby Boomers were born after World War II into an analogue, print-dominated age, when the Cold War and space missions were high on the political agenda and civil rights issues and anti-establishment attitudes prevailed. The internet was in its infancy and was the preserve of academics and scientists as a means of rapid scholarly communication. In 1945, Vannevar Bush outlined his idea of the memex (Bush, 1945), envisaging a mechanical system of providing access to the world's growing scientific literature as well as personal notes and visual material such as photographs. This vision provided a foundation for pioneers such as Douglas Engelbart and Ted Nelson to build hypertext systems in the 1960s, and for Tim Berners-Lee, himself a Baby Boomer, to establish the hypertext protocol for the World Wide Web in 1989.

Tapscott's Bust generation, also known as Generation X, grew up in a hybrid technological age, experiencing the rapid development of home computing, mobile communications, the emergence to the general population of the WWW in the early 1990s, and political events such as the fall of the Berlin Wall, and the Tiananmen Square protests. This generation has experienced the rapid transition from print to electronic information and has adapted to and embraced the online world, while at the same time sharing the values of the Baby Boomers.

The Net Generation—also known as Millennials, Generation Y, and the Google Generation—are "technology veterans," having been born into a fully digital world. Digital natives, they have grown up with computers, search engines and electronic games, using the internet for school, work and leisure and multitasking by interacting naturally with social technologies such as instant messaging, blogs, wikis and Web 2.0 functionality which facilitates collaboration and information sharing. Demanding and impatient, they have high expectations of technology.

Early studies into information-seeking behavior reflect the emerging technologies available to Baby Boomers and Generation X students and focus on the use and design of information systems such as online library catalogues (Borgman, 1983, 1986a; Mitev, Venner, & Walker, 1985; Markey, 1984; Matthews, 1982; Hildreth, 1987). A considerable body of research has also investigated cognitive models of interaction with digital information resources using hypertext navigation to emphasize semantic relationships (Borgman, 1986b; Belkin, 1984; Bovey & Brown, 1987; Marchioni & Schneiderman, 1988). During the 1980s, issues ranged from retrieval problems due to impoverished data, which had been converted from printed sources that lacked description and were not designed to accommodate keyword searching, to human-computer interaction studies concerned with interface design, system navigability, user expectations and intuitive usability which required minimal learning effort.

Later research focused on the transition from print to electronic information resources in an increasingly digital environment influenced by the development of electronic networks to support the growth in desktop access to the internet, to bibliographic and full text databases, and to the use of email (Adams & Bonk, 1995; Budd & Connaway, 1997). These large electronic information resources aimed to make the world's body of published knowledge accessible to the academic community, and research at this time reflected a growing interest in information-seeking behavior. Indeed, trends in scholarly information-seeking behavior between 1995 and 2000 show that younger, X-Generation academics were making greater use of electronic resources (Herman, 2001; Lazinger et al., 1997; Milne, 1999; Zhang, 2001; Whitmire, 2001). The use of electronic scholarly journals by the academic community has been examined in detail by Tenopir (2003), who found disciplinary as well as age-related differences in the usage of electronic resources.

Search engine transaction logs have yielded much interesting data about query formulation and searching behavior on the internet by the public as well as by academics. Studies at CIBER (the Centre for Interactive Behaviour and the Evaluation of Research) at University College London have analyzed transaction logs to map user behavior in large, full text databases finding that users tend to carry out short searches, which they rarely modify, and view few web pages resulting from their search. Usage patterns of search engines also demonstrate shallow, promiscuous or indiscriminate and dynamic forms of behavior indicating limited site penetration, with users visiting many sites without returning to them. (Spink et al., 2001; Jansen & Spink, 2006; Nicholas, Huntington, & Watkinson, 2003; Nicholas et al., 2004; Nicholas et al., 2005; Nicholas et al., 2006; Jamali, Nicholas, & Huntington, 2005). This chaotic behavior reflects the seemingly haphazard activities of the impatient, collaborative Net Generation, perhaps influenced by the educational paradigm shift from traditional broadcast, teacher-dominated instruction to interactive, problem and resource-based participative learning (Fidel et al. 1999). Information-seeking behavior has also been related to personality traits and motivation (Heinstrom, 2003, 2006), and evidence that users have little understanding of how search engines work adds to the confusion (Pew Internet and American Life Project, 2005; Muramatsu & Pratt, 2001).

Focus on the End User

In terms of being information wise, concern is growing about the effectiveness of end user searching, and questions are being raised about the ability of internet users—in particular, the Net Generation—to be able to recognize and filter out misleading or poor quality information (Oblinger & Hawkins, 2006; Lorenzo & Dziuban, 2006; Brown, Murphy & Nanny, 2003). Whilst the Net Generation is perceived to be technologically competent, digitally literate and information savvy in that they are confident that all the information they need can be found on the internet, doubts are surfacing about their information-seeking behavior and the ability of Millennial students to apply relevancy tests to the information they find and their understanding of the legal and ethical considerations of copyright and intellectual property as well as the critical thinking skills that underpin information literacy (Rogers & Swan, 2004).

Plagiarism, or "the practice of claiming, or implying, original authorship of (or incorporating material from) someone else's written or creative work, in whole or in part, into one's own without adequate acknowledgement" (Wikipedia, 2007, no page) has become a major issue at all levels of education. With increasingly ready access to vast quantities of information and a lack of awareness about copyright and intellectual property law, there is a growing trend for many students to copy information from the internet and use it as a means to an end, without considering the meaning and sense of the work they are producing (Williamson & McGregor, 2006; Carroll, 2002; Ercegovac, 2005). Numerous conferences and websites, such as the JISC Plagiarism Advisory Service (see http://www.jiscpas.ac.uk) and Plagiarism.org are devoted to the topic, while thousands of institutions worldwide make use of the plagiarism detection software, Turnitin (see http://www.turnitin.com) to address the problem.

While it can be assumed that the Net Generation is the most likely to be digitally literate and information savvy, it is interesting to compare information-seeking behavior across a range of age groups. Much research has focused on scholarly use of electronic information sources by students, researchers and faculty (e.g., Herman, 2001; Tenopir, 2003). Although heavy users of digital resources, the motivation of these groups to seek information for academic purposes and the specialist nature of subscription information retrieval systems may lead them to display atypical internet searching behavior. Studies of children's use of the internet show that they prefer visual information and encounter problems in retrieving information and revising search strategies. Reasons for this include immature problem-solving skills, limited vocabulary and subject knowledge, and incomplete conceptual models of the internet, which tend to develop with age and experience. Narrow approaches to learning, encouraged by curriculum constraints, and teachers' own lack of experience are also contributory factors to children's ineffective information retrieval and revision strategies (Fidel, 1999; Large, 2004; Bilal & Kirby, 2002; Madden et al., 2006; Slone, 2003).

Information-seeking behavior by the public is dominated by queries about pop culture, news, and events, and search engine results are generally trusted, despite the fact that they frequently return sponsored links and advertisements, though these are not always recognized as such (Pew Internet and American Life Project, 2005). Digital natives are enthusiastic searchers and completely at ease in an electronic environment; they are confident about their ability to find information. Nonetheless, surveys tell us that older internet users—or "digital immigrants"—quickly develop skills to a similar level of expertise. Many Baby Boomers and those over 65 are engaging in online activity, often to communicate with family members. Experience with digital technology at work means that 50–64-year-olds use the internet as much as other groups and those aged over 65 years spend an average 42 hours a month on the internet, compared with teenagers' average of 25 hours per month (Pew Internet and American Life Project, 2001; Ofcom, 2006).

The Generation Gap

The digital generation gap represents something of a dichotomy, with digital natives and digital immigrants using different languages. With no experience of pre-digital life, members of the Net Generation do not describe things in terms of them being digital, since they always have been, so there is no alternative. To them, computers are not technology, they are part of life. Digital immigrants, on the other hand, speak a language which reflects their experience of pre-digital life, by describing things as "digital" to differentiate between electronic and traditional versions. Digital immigrants do, though, speak different dialects, according to the degree of their immersion in the digital world and level of information savviness, with some "speaking digital" more naturally than others. That being said, however, for the next five to ten years, until Baby Boomers retire from the workforce, digital immigrants will dominate as teachers, professors and employers, and linguistic incompatibilities will increase the potential for misunderstandings and communication problems. (Gartner, 2007; Oblinger, 2003)

The literature suggests that learning styles are another example of the generation gap. For example, traditionally structured academic tasks are being replaced by a resource-dependent, problem-based, interactive educational culture; the "guide on the side" is taking over from the "sage on the stage." Digital natives like instant information, prefer graphics, animations, audio, and video to text, and naturally interact with others while multitasking. For them, *doing* is more important than knowing, and learning has to be fun and instantly relevant. Digital immigrants prefer to handle knowledge systematically, logically and to inform discrete activities. Part of an aging infrastructure, many academics display a reluctance to use technology in the classroom and tend to rely on email to communicate with students, rather than using chat, instant messaging, and discussion boards. Inconsistencies in the availability and standard of internet access in many schools seem to frustrate efforts to integrate web technology into the curriculum and the road to information wisdom may be hindered by students who consider themselves more internet savvy than their teachers (Oblinger, 2003; Pew Internet and American Life Project, 2002a, 2002b; Selwyn, 2006; Gardner & Eng, 2005; Carlson, 2005).

Instructional technologies such as Blackboard and Moodle allow access to course materials, assignment grades and discussion fora and are making inroads into teaching practice. Electronic, or virtual learning environments, also known as learning—or course—management systems are used to provide course material for distance learners and to complement face-to-face teaching. Indeed, as early as 1998, Tapscott identified computer-aided instruction (CAI) systems as a means of improving learning (1998, p. 140), lauding their potential to become the digital learning environment of the future. In the U.K., the Higher Education Funding Council announced a 10-year strategy to embed virtual learning environments in universities using technology to transform higher education into a student-focused, flexible system to support lifelong learning (HEFCE, 2005). Such courseware supports the current constructivist theory of interactive learning as a means of creating meaning and developing individual models of knowledge. It also offers opportunities to embed information literacy tuition in teaching through collaboration between librarians and academics. Numerous examples of online tutorials, which can be easily integrated into virtual learning environments, have been developed such as TILT (University of Texas; see http://tilt.lib.utsystem.edu/), PILOT (Queensland University of Technology; see http://pilot.library.qut.edu.au/) and Safari at the UK's Open University (see http://open.ac.uk/safari). Information wisdom is most successfully attained when information literacy tuition is delivered as part of the curriculum, to provide examples relevant to real information needs.

As part of the information universe, information literacy has a continuing role in terms of lifelong learning and the workplace. The concept of knowledge management as a means of harnessing information and expertise in a commercial context, which can be shared among employees and used to the benefit of companies and organizations, facilitates information handling and coexists happily with the notions of lifelong learning and information literacy. However business leaders and technology transfer specialists tend to consider information handling in terms of ICT competency, rather than in terms of being information literate (cf., Cheuk, 2002; Klingner & Sabet, 2005).

The Information-Seeking and -Handling Revolution

Earlier in the chapter two important points made by Gilster were identified as critical to the concept of being information savvy. The first suggests that the digital environment has revolutionized not only information-seeking, but also information-handling behavior.

This revolution is confirmed by more than five years of work investigating the information-seeking behavior of hundreds of thousands of users from a whole range of virtual environments (e.g., health, media, and publishing) by CIBER at University College London. It has raised considerable concerns as to how (and whether) virtual information consumers are coping with mass information on-tap; whether they are really performing and benefiting as they should be or as educationists, librarians and policy makers would have expected them to. Many of the behavioral patterns identified would suggest that: (1) a "dumbing" down in information seeking has taken place as a result of disintermediation, a wholesale reliance on search engines and bewildering choice; (2) that the vast numbers of newly information-enfranchised people who have been introduced to digital information retrieval courtesy of the internet are having all kinds of difficulties coping. After all, many of these people would not have had contact with a library or database before or have obtained any information training; the web is their first real information experience, and they are patently unprepared for the riches and complications they find. It is rather like learner drivers being given a ten-lane motorway to train on.

It is possible that we are witnessing failure at the terminal on a truly astonishing scale, and nobody seems to care because information seeking has been completely disintermediated and is an activity which has become commercially attractive. The default position is that with unprecedented levels of information access, it is up to you to make the most of it.

What is really troubling about this situation is that we are not just talking about people trying to find books, because increasingly people are being forced to use the Web just to be able to function in today's society. Health, wealth and education are now dependent on access to the internet and the ability to use it effectively. Many opportunities and much information are only offered to the virtual consumer.

What then are the characteristics of the information seeking-behavior that are causing concern? In broad terms this behavior can be portrayed as being "bouncing" and viewing in nature. It is also promiscuous and volatile. Bouncing is a form of behavior where users view only one or two web pages from the vast numbers available to them, and a substantial proportion (usually the same ones) generally do not return to the same website very often, if at all. Bouncing can be construed to point to negative outcomes (not finding what you want, short attention spans, etc.), as does another piece of evidence, that concerning online viewing—on average most people spend only a few minutes on a visit to a website, insufficient time to do much reading or obtain much understanding. When put together with the bouncing data it would appear that we are witnessing the emergence of a new form of "reading" with users "power browsing" horizontally through sites, titles, contents pages and abstracts, going for quick wins. It almost seems that they go online to avoid reading in the traditional sense. So it seems that much of the tremendous amount of activity that we see on the web does not really constitute use or satisfaction and, perhaps, represents people trying to find things and not succeeding very often.

Anecdotal evidence supports what CIBER has found in its logs of internet searches. The message coming from libraries, schools and academics is that young, information-savvy people think that finding information is simple; also they *want* simple information, served up in bite-sized chunks. Fast food and obesity come to mind when thinking of analogous situations.

Can anyone put this right and if so who? The assumption generally is that it really has to be teachers and librarians; we surely cannot leave it to intelligent systems, most of which have a hidden financial agenda. However, the omens are not good here. Firstly, all the evidence points to the fact that users do not believe they need information literacy training; they tend to confuse easy access with easy and effective searching. Secondly, the evidence shows that schools are actually doing less training than they did in a pre-digital world where the problems were far fewer and the stakes not so high. Information literacy training appears not to be needed in a disintermediated, "me" environment where huge trust is placed on the abilities of the search engine to second-guess your personal information needs. It is just not cool. If the information literacy message is to be heard then it will only be heard if it can be connected to real world outcomes. The Herculean task for teachers and librarians is to persuade information consumers that if they really do take on board information literacy (i.e., learn about the sources, what is authoritative and what isn't), then they will get a better qualification, degree, job, or life-outcome. There has to be some kind of pay-back; that's why the information profession in particular desperately needs to come up with outcomes data; hard data which demonstrate conclusively that, say, if you attend this literacy program, if you really search the library's databases, and don't just use Google, it will make a difference and you will end up with a higher grade, better degree, etc.

From Information Savvy to Information Literacy: Educating for Discrimination in the Digital World

The second point made by Gilster in *Digital Literacy*, identified earlier in this chapter as critical to the concept of being information savvy, is that technical skills may be less important than a discriminating view of what is found on the internet.

Having considered the interdependencies between digital and information literacies in some detail, it is clear that the process of finding information in a digital environment requires very different skills to those needed to judge its quality and value. While we are digitally literate enough to be able to use a web browser and a mouse to click on links and navigate our way through the e-mass of the internet, and information savvy enough to be able to search for and locate information, how information wise have we become in terms of evaluating the content we find there? Do we understand its meaning? Does it fully meet our needs? How do we use it? Do we know whether it is reliable, truthful or impartial? Is the failure to locate everything that is relevant in this e-mass a problem?

Before looking at how information literacy can enable us to achieve information wisdom, two important aspects of being information savvy have to be mentioned:

- the internet equivalent of being "street-wise"
- the inaccessibility of those parts of the internet's structure that search engines cannot reach.

Information savviness is about creating information as well as finding it, and there is a need for awareness of internet etiquette. Being street wise means exercising caution in posting personal information on your Facebook or MySpace profile, where it can be viewed by not only friends, but also parents, future employers or even predators. Blogs and wikis are useful tools for recording information and creating personal content, but their very visibility and accessibility make their creators vulnerable to exposure.

While the quantity of information that can be retrieved from the internet is vast, it represents only the proportion that is accessible by search engines and is identified by robots or web crawlers. These "agents" visit websites and index documents and pages from title words or other tagged data and can also follow links to identify related documents. The indexing process, however, is far from exhaustive; many web pages and documents are part of the "invisible" or "deep" web, which comprises access-controlled and dynamically generated pages. Those who rely on Google or Yahoo to find information to solve problems or answer questions are thus failing to access a significant proportion of relevant information. The internet and its associated search engines, which also work in different ways, are merely enablers which allow us to locate readily available and easily visible content. Although on the one hand much valuable information on the internet is inaccessible, on the other, search engines are increasingly returning links to perhaps less desirable or useful websites. The ever-increasing presence of sponsored links and advertisements adds to users' confusion and challenges our ability to understand and use information in an appropriate and ethical manner.

But, if information-savvy users are confident in their ability to use the internet to locate information, can information literacy tuition change their information-seeking behavior by encouraging them to acquire and apply the critical and analytical skills that will support the continuous learning process and make them information wise? Being able to discriminate between what is reputable and what is misleading or subversive, and to use information ethically, respecting intellectual property and copyright is necessary to the process and to meet the internationally defined standards for information literacy that were described earlier.

Typically, Net Generation students learn independently and seek advice from peers rather than librarians. Learning has become a lifelong commitment and being information literate is critical to social inclusion. The information haves are more capable of functioning effectively in a digital society than the have-nots and formal information literacy education is one means of developing information savviness. As the Australian and New Zealand Institute for Information Literacy standards (Bundy, 2004, p. 4) state, information literacy is a prerequisite for:

- "participative citizenship
- social inclusion
- the creation of new knowledge
- personal, vocational, corporate and organisational empowerment
- learning for life"

Attaining information literacy is essential for lifelong learning. Formal education is a key route to developing the necessary skills, particularly at university level, although work is also being done in schools. Digital literacy, together with a cognitive awareness of the e-space of the internet, is a major influence on information-seeking behavior and the effective satisfaction of information needs. Such needs are increasingly associated with academic, workplace, health, legal and cultural activities or social and leisure pursuits. Lifelong learning is essential in the workplace, and employers will increasingly depend on an information-wise workforce to function efficiently in an information-rich business environment. While many employees are skilled at using software applications, and can build spreadsheets, use email, surf the internet, and word process, these technological skills are the stuff of digital literacy. The ability to determine what information is needed and locate, evaluate and manage information distinguishes the information wise from the merely savvy and eliminates potentially costly risks to businesses.

Information professionals are concerned with the quality of information, whilst for end users information that is good enough will often suffice. The gap between these objectives is one that librarians strive to fill by encouraging users to think critically about what they find. Library staff have historically influenced information-seeking behavior by acting as intermediaries and educating and advising users on how to achieve the best results from library catalogues, literature searching and subscription databases, by means of bibliographic instruction, user education or, more recently, information literacy programs. These can vary from self-directed online tutorials which guide users through the range of information resources that are available such as library catalogues and electronic journals to problem solving approaches and taught programs that are developed collaboratively with academic staff and delivered as part of the curriculum.

The Future for Information Literacy

If information-seeking behavior, which reflects changing learning styles and unprecedented access to vast quantities of information, is typically haphazard and chaotic it would seem desirable that some sort of order be imposed that will enable users to apply the critical thinking skills they apparently need to evaluate and interpret the information retrieved and become information literate. Information literacy is a fundamental part of the learning process, just as digital literacy underpins the ability to use technology effectively.

Professional librarians are attempting to address this by being actively involved in promoting information literacy at all levels of education. Evidence suggests that information-seeking behavior can be transformed where there is a genuine information need and the processes of evaluation are set in a subject context. However, hard data which demonstrate that information literacy makes a difference are scarce, and support for the integration of initiatives into school and higher education curricula is only slowly forthcoming. The goal of developing an information-literate population equipped to participate in the creation of new knowledge and able to deal with the economic, social, legal and ethical issues of seeking and handling information seems to be elusive. Without high-level, strategic direction these issues will remain wide open.

One problem is that librarians are not, by and large, trained to teach, which raises questions about pedagogic credibility in the quest to embed information literacy in the educational process. Information literacy is an education-wide responsibility, and, if information-seeking behavior is to be influenced to the extent that information wisdom can be attained, collaboration between librarians, academics, administrators and learning technologists is needed to integrate programs into course content and establish pedagogically sound assessment techniques. Information literacy has to be guided by learning theory if the crossover between bibliographic instruction and teaching practice is to occur.

There is also a generational barrier to accepting the potential of information literacy, in that education systems are currently dominated by Baby Boomers, who do not always recognize the need to update their levels of digital literacy since they can get by on the skills they have already acquired even if they are not the most efficient or effective. The lack of recognition of the extent to which digital literacy infiltrates every aspect of life leaves more aware students and researchers confused by unfamiliar and outdated practices and methods.

Hard evidence of the successful outcomes of information literacy programs is needed if users and those in authority such as policy makers and educational strategists are to be convinced of the deficiencies of being merely information savvy in an information-dependent but complex and bewildering digital society. The goal of attaining information wisdom through digital and information literacy needs to be set at a young age and persist at all educational levels as part of a reward system if information-seeking behavior is to be influenced and information-literate individuals are to be delivered into the workplace and beyond.

References

Adams, J. & Bonk S. (1995). Electronic information technologies and resources: use by university faculty and faculty preferences for related services. *College and Research Libraries.* 56(2), 119–131.

American Library Association (1989). *Presidential committee on information literacy: Final report.* Washington, DC: American Library Association. Available from: http://www.ala.org/ala/acrl/acrlpubs/whitepapers/presidential.htm (accessed 21 Sept 2007).

——— (2001). *ACRL competency standards for information literacy in higher education.* Washington, DC: American Library Association. Available from: http://www.ala.org/ala/acrl/acrlstandards/informationliteracycompetency.cfm (accessed 21 Sept 2007)

Bawden, D. (2001). Information and digital literacies: A review of concepts. *Journal of Documentation,* 57(2), 218–259.

Belkin N. (1984). Cognitive models and information transfer. *Social Science Information Studies.* 4, 111–129.

Bilal, D. & Kirby, J. (2002). Differences and similarities in information seeking: Children and adults as Web users. *Information Processing and Management.* 38, 649–670.

Borgman, C. (1983). *End user behaviour in the Ohio State Universities' online catalog: A computer monitoring study.* Dublin, OH: OCLC Online Computer Library Center.

———. (1986a). Why are online catalogues hard to use? Lessons learned from information retrieval studies. *Journal of the American Society for Information Science and Technology.* 37(6), 387–400.

———. (1986b). The user's mental model of an information retrieval system: An experiment on a prototype online catalogue. *International Journal of Man-Machine Studies.* 24(1), 47–64.

Bovey, J. & Brown, P. (1987). Interactive document display and its use in information retrieval. *Journal of Documentation.* 43(2), 124–137.

Brown, C., Murphy, T., & Nanny, M. (2003). Turning techno-savvy into info-savvy: Authentically integrating information literacy into the college curriculum. *The Journal of Academic Librarianship.* 29(6), 386–398.

Budd, J. & Connaway, L. (1997). University faculty and networked information: Results of a survey. *Journal of the American Society for Information Science.* 48(9), 843–852.

Bundy A. (Ed.) (2004). *Australian and New Zealand information literacy framework: Principles, standards and practice.* (2nd edn). Adelaide: Australian and New Zealand Institute for Information Literacy. Available from: http://www.caul.edu.au/info-literacy/InfoLiteracyFramework.pdf (accessed October 2007).

Bush, V. (1945). As we may think. *Atlantic Monthly,* July, 101–108.

Carlson, S. (2005). The net generation goes to college. *Chronicle of Higher Education.* 52, 7th October.

Carroll, J. (2002). *Handbook for deterring plagiarism in higher education.* Oxford: Oxford Brookes University.

CAUL (Council of Australian University Librarians) (2001). Information literacy standards. Available from: http://www.caul.edu.au/caul-doc/InfoLitStandards2001.doc (accessed October 2007).

Cheuk, B. (2002). Information literacy in the workplace context: Issues, best practice and challenges. White Paper prepared for UNESCO, the US National Commission on Libraries and Information Science and the National Forum on Information Literacy, for use at the Information Literacy Meeting of Experts, Prague. Available from: http://www.nclis.gov/libinter/infolitconf&meet/papers/cheuk-fullpaper.pdf (accessed 8 October 2007).

CILIP (Chartered Institute of Library and Information Professionals) (2004). Information literacy definition. Available from: http://www.cilip.org.uk/professionalguidance/informationliteracy/definition/ (accessed 21 Sept 2007).

Devlin, B. (1997). Conceptual models for network literacy. *The Electronic Library*. 15(5), 363–368.

DigEuLit Project (2005–2007). Available from: http://www.digeulit.ec/ (accessed 21 Sept 2007).

Ercegovac, Z. (2005). What do students say they KNOW, FEEL and DO about cyber-plagiarism and academic dishonesty? A case study. *Proceedings of the American Society for Information Science and Technology*. 42(1). Available from: http://eprints.rclis.org/archive/00005063/ (accessed October 2007).

Erstad, O. (2006). A new direction? Digital literacy, student participation and curriculum reform in Norway. *Education and Information Technologies*. 11(3), 415–429.

Fidel, R., Davies, R., Douglas, M., Holder J., Hopkins, C., Kushner, E., Miyagishma, B., & Toney, C. (1999). A visit to the information mall: Web searching behavior of high school students. *Journal of the American Society for Information Science*. 50(1), 24–37.

Gardner, S. & Eng, S. (2005). What students want: Generation Y and the changing function of the academic library. *Libraries and the Academy*. 5(3), 405–420.

Gartner (2007). Digital Natives hit the workplace: from students to employees. Gartner Symposium, IT^xpo, Barcelona 20–23 May. Available from: http://agendabuilder.gartner.com/spr8/WebPages/SessionDetail.aspx?EventSessionId=900 (accessed October 2007).

Gilster, P. (1997). *Digital literacy*. John Wiley, 1997

HEFCE (Higher Education Funding Council for England) (2005) HEFCE strategy for e-learning. Available from: http://www.hefce.ac.uk/pubs/HEFCE/2005/05_12/ (accessed October 2007).

Herman, E. (2001). End users in academia: Meeting the information needs of university researchers in an electronic age. Part 2. *Aslib Proceedings*. 533(10), 431- 457.

Heinstrom, J. (2003). Fast surfers, broad scanners and deep divers as users of information technology—relating information preferences to personality traits. Paper presented to the American Society for Information Science and Technology (ASIST) Annual Meeting: Humanizing Information Technology: From Ideas to Bits and Back. Long Beach, California, October 20—23.

———. (2006). Fast surfing for availability or deep diving into quality: Motivation and information seeking among middle and high school students. *Information Research*. 11(4). Available from: http://informationr.net/ir/11–4/paper265.html (accessed October 2007).

High-Level Colloquium on Information Literacy and Lifelong Learning (2005). IFLA World Summit on the Information Society. Available from: http://www.ifla.org/III/wsis/info-lit-for-all.htm (Accessed 26 April 2007).

Hildreth, C. (1982). *Online public access catalogues: The user interface*. Dublin, OH: OCLC Online Computer Library Center.

———. (1987). Beyond Boolean: Designing the next generation of library catalogues. *Library*

Trends. 35(4), 647–667.

Jamali, H., Nicholas, D., & Huntington P. (2005). The use and users of scholarly e-journals: a review of log analysis studies. *Aslib Proceedings.* 57(6), 554–571.

Jansen, B. & Spink, A. (2006). How are we searching the World Wide Web? A comparison of nine search engine transaction logs. *Information Processing and Management.* 42(1), 248–263.

Klingner, D. & Sabet, G. (2005). Knowledge management, organizational learning, innovation, and technology transfer: What they mean and why they matter. *Comparative Technology Transfer and Society.* 3(3), 199–210.

Large, A. (2004). Children, teenagers and the web. *Annual Review of Information Science and Technology.* (39), 347–392.

Lazinger, S., Bar-Ilan, J., & Peritz, B. (1997). Internet use by faculty members in various disciplines: A comparative case study. *Journal of the American Society for Information Science.* 48(6), 508–518.

Lorenzo, G. & Dziuban, C. (2006). Ensuring the net generation is net savvy. ELI Paper 2. *Educause Learning Initiative.* Available from: http://www.educause.edu/ir/library/pdf/ELI3006.pdf (accessed 23.07.2007)

Madden, A., Ford, N., Miller, D., & Levy, P. (2006). Children's use of the internet for information seeking. *Journal of Documentation.* 62(6), 744- 761.

Marchioni, G. & Schneiderman, B. (1988). Finding facts vs. browsing knowledge in hypertext systems. *IEEE Computer.* 21(1), 71–80.

Markey, K. (1984). *Subject searching in library catalogues.* Dublin, OH: OCLC Online Computer Library Center.

Martin, A. (2006). A framework for digital literacy, DigEuLit Project working paper. Available from: http://www.digeulit.ec/docs/public.asp?id=3334 (accessed 21 Sept 2007).

———. (2005). DigEuLit—a European framework for digital literacy: A progress report. *Journal of eLiteracy.* 2(2), 130–136. Available from: http://www.jelit.org/65/01/JeLit_Paper_31.pdf (accessed 21 Sept 2007).

Martin, A. & Madigan, D. (Eds.) (2006). *Digital literacies for learning.* London: Facet Publishing.

Matthews, J. (1982). *A study of six online public access catalogs: A review of findings.* Report submitted to the Council on Library Resources. Grass Valley, CA: J. Matthews & Associates.

McGregor, J. & Streitenberger, D. (1998). Do scribes learn? Copying and information use. *School Library Media Quarterly Online.* 1(1). Available from: http://www.ala.org/ala/aasl/aaslpubsandjournals/slmrb/slmrcontents/volume11998slmqo/mcgregor.htm (accessed 21 September 2007).

Mikhailova A. (2006). By day she worked on Harry Potter, but by night . . . Revealed: identity of erotic diarist behind summer's hottest book. *Sunday Times,* 6 August, 2006. Available from: http://www.timesonline.co.uk/tol/news/uk/article601445.ece (accessed 21 September 2007).

Milne, P. (1999). Electronic access to information and its impact on scholarly communication. *Proceedings of the Conference on Information Online and on Disc 99.* January 19–21, Sydney, NSW, Australia. Available from: http://www.csu.edu.au/special/online99/proceedings99/305b.htm (accessed 9 July 2007).

Mitev, N., Venner, G., & Walker, S. (1985). *Designing an online public access catalogue.* London:

The British Library.

Muramatsu, J., & Pratt, W. (2001). Abstract: Transparent queries: Investigating users' mental models of search engines. *ACM SIGIR*, New Orleans, 9–12 Sept. Available from: http://portal.acm.org/citation.cfm?id=383991&dl=GUIDE. (Accessed 27 April 2007).

Navarro-Prieto, R., Scaife, M., & Rogers, Y. (2007). Cognitive strategies in web searching. Available from: http://zing.ncsl.nist.gov/hfweb/proceedings/navarro-prieto/ (accessed 27 April 2007).

Nicholas, D., Huntington, P., & Watkinson, A. (2003). Digital journals, big deals and online searching behaviour: A pilot study. *Aslib Proceedings.* 55(1/2), 84–109.

Nicholas, D., Huntington, P., Williams, P., & Dobrowolski, T. (2004). Re-appraising information seeking behaviour in a digital environment: Bouncers, checkers, returnees and the like. *Journal of Documentation*, 60(1), 24–43.

Nicholas, D., Huntington, P., & Watkinson, A. (2005). Scholarly journal usage: The results of deep log analysis. *Journal of Documentation.* 61(2), 248–280.

Nicholas, D., Huntington, P., Jamali, H., & Watkinson, A. (2006). The information seeking behaviour of the users of digital scholarly journals. *Information Processing & Management.* 42(5), 1345–1365.

Nicholas, D., & Williams, P. (1998). *Digital Literacy*, by Paul Gilster (Book review) *Journal of Documentation.* 54(3), 360–362.

Oblinger, D. (2003). Boomers, Gen Xers and Millenials: Understanding the new students. *Educause Review.* 38, July/August. Available from: http://www.educause.edu/ir/library/pdf/erm0342.pdf (accessed October 2003).

Oblinger, D. & Hawkins, B. (2006). The myth about student competency: Our students are technically competent. *Educause Review,* 41(2),12–13.

Ofcom (2006). *Telecommunications in the communications market 2006.* London: Office of Communication. Available from: http://www.ofcom.org.uk/research/cm/cm06/cmr06_print/telec.pdf (accessed October 2007).

Pew Internet and American Life Project (2001). *Wired seniors.* Report by the Pew Internet and American Life Project. Available from: http://www.pewinternet.org/PPF/r/40/report_display.asp (accessed October 2007).

——— (2002a). *The internet goes to college: How students are living in the future with today's technology.* Report by the Pew Internet and American Life Project. Available from: http://www.pewinternet.org/PPF/r/71/report_display.asp (accessed October 2007).

——— (2002b) *The digital disconnect: The widening gap between internet savvy students and their schools.* Report by the Pew Internet and American Life Project. Available from: http://www.pewinternet.org/PPF/r/67/report_display.asp (accessed October 2007).

——— (2005) *Search engine users.* Report by the Pew Internet and American Life Project. Available from: http://www.pewinternet.org/PPF/r/146/report_display.asp (accessed October 2007).

Rogers, D. & Swan, K. (2004). Self regulated learning and the internet. *Teachers College Record.* 106(9), 1804–1824.

Schneiderman, B. (1986). *Designing the user interface: Strategies for effective human computer interaction.* Reading, MA: Addison-Wesley.

SCONUL (Society of College, National and University Libraries) (1999). *Information skills in higher education.* A position paper prepared by the Information Skills Task Force, Society of College, National and University Libraries. Available from: http://www.sconul.ac.uk/groups/information_literacy/papers/Seven_pillars.html (accessed October, 2007).

Selwyn, N. (2006). Exploring the 'digital disconnect' between net-savvy students and their schools. *Learning Media and Technology.* 31(1), 5–17.

Slone, D. (2003). Internet search approaches: The influence of age, search goals, and experience. *Library and Information Science Research.* 25(4), 403-418.

Spink, A., Wolframm D., Jansen, B., & Saracevic, T. (2001). Searching the web: The public and their queries. *Journal of the American Society for Information Science and Technology.* 52(3), 226–234.

Tapscott, D. (1998). *Growing up digital.* New York: McGraw-Hill.

Tenopir, C. (2003). *Use and users of electronic library resources: An overview and analysis of recent research studies.* Washington, DC: Council on Library and Information Resources. Available from: http://www.clir.org/pubs/abstract/pub120abst.html (accessed 21 Sep 2007).

Turnitin (2007). Available from: http://www.turnitin.com/static/home.html (accessed 21 Sept 2007).

Whitmire, E. (2001). A longitudinal study of undergraduates' academic library experiences. *Journal of Academic Librarianship.* 27(5), 378–385.

Wikipedia (2007). Plagiarism entry. Available from: http://en.wikipedia.org/wiki/Plagiarism (Accessed 21 Sept 2007).

Williams, D. & Wavell, C. (2006). *Information literacy in the classroom: Secondary school teachers' conceptions.* Research Report 15. Aberdeen: Department of Information Management, Aberdeen Business School, Robert Gordon University. Available from: http://www.rgu.ac.uk/files/ACF4DAA.pdf (accessed October 2007).

Williamson, K. & McGregor, J. (2006). Information use and secondary school students: A model for understanding plagiarism. *Information Research*, 12(1), paper 288. Available from: http://InformationR.net/ir/12–1/paper288.html (Accessed 21 Sept 2007).

Zhang, Y. (2001). Scholarly use of Internet-based electronic resources. *Journal of the American Society for Information Science.* 52(8), 628–654.

CHAPTER FOUR

Defining Digital Literacy

What Do Young People Need to Know About Digital Media?

DAVID BUCKINGHAM

If you want to use television to teach somebody,
you must first teach them how to use television.
(Umberto Eco, 1979)

Introduction

Umberto Eco's argument about the educational use of television can equally be applied to newer media. As Eco implies, media should not be regarded merely as teaching aids or tools for learning. Education *about* the media should be seen as an indispensable prerequisite for education *with* or *through* the media. Likewise, if we want to use the internet or computer games or other digital media to teach, we need to equip students to understand and to critique these media: we cannot regard them simply as neutral means of delivering information, and we should not use them in a merely functional or instrumental way.

My aim in this chapter is to identify some of the forms that this education might take and some of the questions that it might raise. I argue for a particular definition of "digital literacy" that goes well beyond some of the approaches that are currently adopted in the field of information technology in education. Indeed, implicit in my argument is a view that new digital media can no longer be regarded simply as a matter of "information" or of "technology." This is particularly the case if we are seeking to develop more effective connections between children's experiences of technology outside school and their experiences in the classroom.

With the growing convergence of media (which is driven by commercial forces as much as by technology), the boundaries between "information" and other media have become increasingly blurred. In most children's leisure-time experiences, computers are much more than devices for information retrieval: they convey images and fantasies, provide opportunities for imaginative self-expression and play, and serve as a medium through which intimate personal relationships are conducted. These media cannot be adequately understood if we persist in regarding them simply as a matter of machines and techniques, or as "hardware" and "software." The internet, computer games, digital video, mobile phones and other contemporary technologies provide new ways of mediating and representing the world and of communicating. Outside school, children are engaging with these media, not as technologies but as *cultural forms*. If educators wish to use these media in schools, they cannot afford to neglect these experiences: on the contrary, they need to provide students with means of understanding them. This is the function of what I am calling digital literacy.

Multiple Literacies

Over the past twenty years, there have been many attempts to extend the notion of literacy beyond its original application to the medium of writing. As long ago as 1986, one of the leading British researchers in the field, Margaret Meek Spencer, introduced the notion of "emergent literacies" in describing young children's media-related play (Spencer, 1986); and the call for attention to "new" or "multiple" literacies has been made by many authors over subsequent years (Bazalgette, 1988; Buckingham, 1993a; Tyner, 1998; and many others). We have seen extended discussions of visual literacy (e.g., Moore & Dwyer, 1994), television literacy (Buckingham, 1993b), cine-literacy (British Film Institute, 2000), and information literacy (Bruce, 1997). Exponents of the so-called New Literacy Studies have developed the notion of "multiliteracies," referring both to the social diversity of contemporary forms of literacy and to the fact that new communications media require new forms of cultural and communicative competence (Cope & Kalantzis, 2000).

This proliferation of literacies may be fashionable, but it raises some significant questions. Popular discussions of "economic literacy," "emotional literacy" and even "spiritual literacy" seem to extend the application of the term to the point where any analogy to its original meaning (that is, in relation to written language) has been lost. "Literacy" comes to be used merely as a vague synonym for "competence," or even "skill." It is worth noting in this respect that such expressions may be specific to the English language.

In some other languages, the equivalent term is more overtly tied to the notion of defining digital literacy writing—as in the French word "alphabetization"; while in other cases, "media literacy" is often translated into a more general term for skill or competence—as in the German "Medienkompetenz."

The term "literacy" clearly carries a degree of social status; and to use it in connection with other, lower status forms such as television, or in relation to newer media, is thus to make an implicit claim for the latter's validity as objects of study. Yet as uses of the term multiply, the polemical value of such a claim—and its power to convince—is bound to decline. Thus, while recognizing the significance of visual and audio-visual media, some scholars challenge this extension of the term, arguing that "literacy" should continue to be confined to the realm of writing (Barton, 1994; Kress, 1997) while others dispute the idea that visual media require a process of cultural learning that is similar to the learning of written language (Messaris, 1994). The analogy between writing and visual or audiovisual media such as television or film may be useful at a general level, but it often falls down when we look more closely: it is possible to analyze broad categories such as narrative and representation across all these media, but it is much harder to sustain more specific analogies, for example, between the film shot and the word, or the film sequence and the sentence (Buckingham, 1989).

Nevertheless, the use of the term "literacy" implies a broader form of education about media that is not restricted to mechanical skills or narrow forms of functional competence. It suggests a more rounded, humanistic conception that is close to the German notion of "Bildung." So what are the possibilities and limitations of the notion of *"digital* literacy"? Is it just a fancy way of talking about how people learn to use digital technologies, or is it something broader than that? Indeed, do we really need yet another literacy?

Towards Digital Literacy

The notion of digital literacy is not new. Indeed, arguments for "computer literacy" date back at least to the 1980s. Yet as Goodson and Mangan (1996) have pointed out, the term "computer literacy" is often poorly defined and delineated, both in terms of its overall aims and in terms of what it actually entails. As they suggest, rationales for computer literacy are often based on dubious assertions about the vocational relevance of computer skills or about the inherent value of learning with computers, which have been widely challenged. In contemporary usage, digital (or computer) literacy often appears to amount to a minimal set of skills that will enable the user to operate effectively with software tools or in performing basic information retrieval tasks. This is essentially a *functional* definition: it specifies the basic skills that are required to undertake particular operations, but it does not go very far beyond this.

For example, the British government has attempted to define and measure the ICT skills of the population alongside traditional literacy and numeracy as part of its *Skills for Life* survey (Williams et al., 2003). This survey defines these skills at two levels. Level 1 includes an understanding of common ICT terminology; the ability to use basic features of software tools such as word-processors and spreadsheets; and the ability to save data, copy and paste, manage files, and standardize formats within documents. Level 2 includes the use of search engines and databases, and the ability to make more advanced use of software tools. In the 2003 survey, over half of the sample of adults was found to be at "entry level or below" (that is, not yet at Level 1) in terms of practical skills. Other research suggests that adults' ability to use search engines for basic information retrieval, for example, is distinctly limited (Livingstone et al., 2005, pp. 23–24).

Another context in which the notion of digital literacy has arisen in recent years is in relation to online safety. For example, the European Commission's "Safer Internet Action Plan" has emphasized the importance of internet literacy as a means for children to protect themselves against harmful content. Alongside the range of hotlines, filters and "awareness nodes," it has funded several educational projects designed to alert children to the dangers of online pedophiles and pornography—although in fact it is notable that many of these projects have adopted a significantly broader conception of internet literacy that goes well beyond the narrow concern with safety. The "Educaunet" materials, for example, provide guidance on evaluating online sources and assessing one's own information needs, as well as recognizing the necessity and the pleasure of risk for young people (see www.educaunet.org).

Even so, most discussions of digital literacy remain primarily preoccupied with *information*—and therefore tend to neglect some of the broader cultural uses of the internet (not least by young people). To a large extent, the concern here is with promoting more efficient uses of the medium—for example, via the development of advanced search skills (or so called "power searching") that will make it easier to locate relevant resources amid the proliferation of online material. Popular guides to digital literacy have begun to address the need to evaluate online content (e.g. Gilster, 1997; Warlick, 2005); yet these formulations still tend to focus on technical "know-how" that is relatively easy to acquire and on skills that are likely to become obsolete fairly rapidly. Much of the discussion appears to assume that information can be assessed simply in terms of its factual accuracy. From this perspective, a digitally literate individual is one who can search efficiently, who compares a range of sources, and sorts authoritative from non-authoritative, and relevant from irrelevant, documents (Livingstone et al., 2005, p. 31). There is little recognition here of the symbolic or persuasive aspects of digital media, of the emotional dimensions of our uses and interpretations of these media, or indeed of aspects of digital media that exceed mere "information."

Bettina Fabos (2004) provides a useful review of such attempts to promote more critical evaluation of online content. In practice, she argues, evaluation "checklists" are often less than effective. Students may feel inadequate assessing sites when they are unfamiliar with the topics they cover and they largely fail to apply these criteria, instead emphasizing speedy access to information and appealing visual design. More to the point, however, such "web evaluation" approaches appear to presume that objective truth will eventually be achieved through a process of diligent evaluation and comparison of sources. They imply that sites can be easily divided into those that are reliable, trustworthy and factual, and those that are biased and should be avoided. In practice, such approaches often discriminate against low-budget sites produced by individuals, and in favor of those whose high-end design features and institutional origins lend them an air of credibility. The alternative, as Fabos suggests, is to recognize that "bias" is unavoidable and that information is inevitably "couched in ideology." Rather than seeking to determine the "true facts," students need to understand "how political, economic, and social context shapes all texts, how all texts can be adapted for different social purposes, and how no text is neutral or necessarily of 'higher quality' than another" (Fabos, 2004, p. 95).

As this implies, digital literacy is much more than a functional matter of learning how to use a computer and a keyboard, or how to do online searches. Of course, it needs to begin with some of the "basics." In relation to the internet, for example, children need to learn how to locate and select material—how to use browsers, hyperlinks and search engines, and so on. But to stop there is to confine digital literacy to a form of instrumental or functional literacy. The skills that children need in relation to digital media are not confined to those of information retrieval. As with print, they also need to be able to evaluate and use information critically if they are to transform it into knowledge. This means asking questions about the sources of that information, the interests of its producers, and the ways in which it represents the world; and understanding how these technological developments are related to broader social, political and economic forces.

Media Literacy Goes Online

This more *critical* notion of literacy has been developed over many years in the field of media education, and in this respect, I would argue that we need to extend approaches developed by media educators to encompass digital media. There are four broad conceptual aspects that are generally regarded as essential components of media literacy (see Buckingham, 2003). While digital media clearly raise new questions and require new methods of investigation, this basic conceptual framework continues to provide a useful means of mapping the field.

Representation

Like all media, digital media represent the world, rather than simply reflect it. They offer particular interpretations and selections of reality, which inevitably embody implicit values and ideologies. Informed users of media need to be able to evaluate the material they encounter, for example, by assessing the motivations of those who created it and by comparing it with other sources, including their own direct experience.

In the case of information texts, this means addressing questions about authority, reliability and bias, and it also necessarily invokes broader questions about whose voices are heard and whose viewpoints are represented and whose are not.

Language

A truly literate individual is able not only to use language but also to understand how it works. This is partly a matter of understanding the "grammar" of particular forms of communication, but it also involves an awareness of the broader codes and conventions of particular genres. This means acquiring analytical skills and a metalanguage for describing how language functions. Digital literacy must therefore involve a systematic awareness of how digital media are constructed and of the unique "rhetorics" of interactive communication: in the case of the web, for example, this would include understanding how sites are designed and structured, and the rhetorical functions of links between sites (cf. Burbules & Callister, 2000, pp. 85–90).

Production

Literacy also involves understanding who is communicating to whom and why. In the context of digital media, young people need to be aware of the growing importance of commercial influences—particularly as these are often invisible to the user. There is a "safety" aspect to this: children need to know when they are being targeted by commercial appeals and how the information they provide can be used by commercial corporations. But digital literacy also involves a broader awareness of the global role of advertising, promotion and sponsorship and how they influence the nature of the information that is available in the first place. Of course, this awareness should also extend to non-commercial sources and interest groups, who are increasingly using the web as a means of persuasion and influence.

Audience

Finally, literacy also involves an awareness of one's own position as an audience (reader or user). This means understanding how media are targeted at audiences and how different audiences use and respond to them. In the case of the internet, this entails an awareness of the ways in which users gain access to sites, how they are addressed and guided (or encouraged to navigate), and how information is gathered about them. It also means recognizing the very diverse ways in which the medium is utilized, for example, by different social groups, and reflecting on how it is used in everyday life—and indeed how it might be used differently. (In some respects, of course, the term "audience" (which is easily applied to "older" media) fails to do justice to the interactivity of the internet—although substitute terms are no more satisfactory (Livingstone, 2004).

Case 1: Web Literacy

How might these broad approaches be applied specifically to studying the World Wide Web? Figure 4.1 indicates some of the issues that might be addressed here and is adapted from Buckingham (2003). It incorporates several of the key concerns of the "web evaluation" approaches discussed above but sets these within a broader context. (Different issues would undoubtedly need to be explored in relation to other uses of the internet, such as email, instant messaging or blogging.)

Figure 4.1: The World Wide Web: Issues for Study.

Representation

- How websites claim to "tell the truth" and establish their authenticity and authority.
- The presence or absence of particular viewpoints or aspects of experience.
- The reliability, veracity and bias of online sources.
- The implicit values or ideologies of web content and the discourses it employs.

Language

- The use of visual and verbal "rhetorics" in the design of websites (for example, graphic design principles, the combination of visuals and text, the use of sound).
- How the hypertextual (linked) structure of websites encourages users to navigate in particular ways.
- How users are addressed: for example, in terms of formality and "user-friendliness".

(*Figure 4.1. Continued on following page.*)

- The kinds of "interactivity" that are on offer and the degrees of control and feedback they afford to the user.

Production

- The nature of web authorship and the use of the internet by companies, individuals or interest groups as a means of persuasion and influence.
- The technologies and software that are used to generate and disseminate material on the web and the professional practices of web "authors."
- The significance of commercial influences and the role of advertising, promotion and sponsorship.
- The commercial relationships between the web and other media such as television and computer games.

Audience

- The ways in which users can be targeted by commercial appeals, both visibly and invisibly.
- The nature of online "participation," from web polls to bulletin boards to "user generated content."
- How the web is used to gather information about consumers.
- How different groups of people use the internet in their daily lives and for what purposes.
- How individuals or groups use and interpret particular sites and the pleasures they gain from using them.
- Public debates about the "effects" of the internet, for example, in relation to online safety and "addiction."

In my view, this approach is significantly more comprehensive and more rigorous than most existing approaches to "internet literacy." It incorporates questions about bias and reliability but sets these within a broader concern with representation. This in turn is related to a systematic analysis of the "grammar" or "rhetoric" of online communications that includes visual as well as verbal dimensions and to an account of the commercial and institutional interests at stake. The approach also entails a reflexive understanding of how these factors impact on the user—how users are targeted and invited to participate, what they actually do with the medium, and what they find meaningful and pleasurable. I would argue that this approach moves well beyond a narrow concern with "information" and a simplistic approach to evaluation that sees it merely in terms of truth and falsity.

Case 2: Game Literacy

The approach outlined here is not only applicable to "information" media. In principle, it can also be applied to other aspects of digital media, including "fictional" media such as computer and video games. Of course, there is a growing interest in using computer games in education, but here again, most proposals implicitly conceive of games as a neutral "teaching aid." In line with Eco's argument about television, I would argue that we also need to be teaching young people about games as a *cultural form*—and that this is a necessary prerequisite for using games in order to teach other curriculum areas.

To date, most proposals for teaching about games in schools have been developed by teachers of English or language arts (e.g., Beavis, 1998). As such, these proposals tend to emphasize the aspects of games that fit most easily with English teachers' traditional literary concerns, for example, with narrative or the construction of character. In terms of our four-part framework, the emphasis is on language and to some extent on representation, but there is little engagement with the more sociological issues to do with production and audience that are important concerns for media teachers.

Equally significantly, this quasi-literary approach can lead to a rather partial account of the textual dimensions of games—which itself raises significant issues about the definition of "game literacy." Clearly, there are many elements that games share with other representational or signifying systems. On one level, this is a manifestation of the convergence that increasingly characterizes contemporary media: games draw upon books and movies, and vice-versa, to the point where the identity of the "original" text is often obscure. Users (players, readers, viewers) must transfer some of their understandings across and between these media, and to this extent it makes sense to talk about "literacies" that operate—and are developed—across media (Mackey, 2002). However, analyzing games simply in terms of these representational dimensions produces at best a partial account. For example, characters in games function both in the traditional way as representations of human (or indeed non-human) "types," and as points of access to the action; but the crucial difference is that they can be manipulated, and in some instances positively changed, by the player. This points to the necessary interpenetration of the *representational* and the *ludic* dimensions of games; that is, the aspects that make games *playable* (Carr et al., 2006).

So is there also a "literacy" that applies to the ludic dimension of games? There is a growing literature, both in the field of game design and in academic research, that seeks to identify basic generative and classificatory principles in this respect (e.g., Salen & Zimmerman, 2003). This kind of analysis focuses on issues such as how games manage time and space, the "economies," goals and obstacles of games, and issues such as rules and conditionality. It is these ludic aspects that distinguish games from movies or books, for example. However, these elements are not separate from, or opposed to, the representational elements, and any account of "game literacy" needs to address *both* the elements that games have in common with other media *and* the elements that are specific to games (whether or not they are played on a computer).

As this implies, the analysis of games requires new and distinctive methods that cannot simply be transferred from other media—although this is equally the case when we compare television and books, for example. While some elements are shared across these media, others are distinctive to a specific medium; and hence we need to talk both in terms of a more general "media literacy" *and* in terms of specific "media literacies" in the plural. Furthermore, developing "game literacy" also needs to address the aspects of production and audience—although here again, the term "audience" seems an inadequate means of describing the interactive nature of play. Figure 4.2 summarizes some of the key issues to be addressed in applying the media literacy framework to computer games and draws on some other recent work in this field (Burn, 2004; Oram & Newman, 2006).

The digital literacy "recipe" outlined here is intended only as a brief indication of the possibilities: more detailed proposals for classroom practice can be found elsewhere (e.g., Burn & Durran, 2007; McDougall, 2006). Obviously, these suggestions will vary according to the needs and interests of the students, although it should be possible to address the general conceptual issues at any level. Nevertheless, it should be apparent that approaching digital media through media education is about much more than simply "accessing" these media or using them as tools for learning: on the contrary, it means developing a much broader *critical understanding*, which addresses the textual characteristics of media alongside their social, economic and cultural implications.

Figure 4.2: Computer Games: Issues for Study.

Representation

- How games lay claim to "realism", for example, in their use of graphics, sounds and verbal language.
- The construction and manipulation of game "characters."
- The representations of specific social groups, for instance, in terms of gender and ethnicity.
- The nature of game worlds and their relationship to real worlds (for example, in terms of history, geography and physics).

Language

- The functions of verbal language (audio and written text), still and moving images, sounds and music.
- The distinctive codes and conventions of different game genres, including the kinds of interactivity—or "playability"—that they offer.
- How different game genres manage space and time (that is, narrative) and how they position the player.
- The ludic dimensions of games—rules, economies, objectives, obstacles, and so on.

Production

- The "authorship" of games, and the distinctive styles of graphic artists and game designers.
- The technologies and software that are used to create games and the professional practices of game companies.
- The commercial structure of the games industry (developers, publishers, marketers) and the role of globalization.
- The relationships between games and other media such as television, books and movies, and the role of franchising and licensing.

(*Figure 4.2. Continued on following page.*)

Audience

- The experience and pleasure of play and how it relates to the rules and structures of games.
- The social and interpersonal nature of play and its functions in everyday life, particularly for different social groups (for example, different genders or age groups).
- The role of advertising, games magazines and online commentary in generating expectations and critical discourse around games.
- Fan culture, including the role of fan websites, fan art, "modding," machinima and so on.
- Public debates about the "effects" of games, for example, in relation to violence.

"Writing" Digital Media

Finally, it is important to recognize that these critical understandings can and should be developed through the experience of media production and not merely through critical analysis. Media literacy involves "writing" the media as well as "reading" them, and here, again, digital technology presents some important new challenges and possibilities. The growing accessibility of this technology means that quite young children can easily produce multimedia texts and even interactive hypermedia—and increasing numbers of children have access to such technology in their homes. Indeed, new media are a key aspect of the much more participatory media culture that is now emerging—in the form of blogging, social networking, game-making, small-scale video production, podcasting, social software, and so on (Jenkins, 2006).

Growing numbers of teachers have sought to harness the productive possibilities of these media, albeit in quite limited ways. As with older media (Lorac & Weiss, 1981), many teachers are using multimedia authoring packages as a means of assisting subject learning in a range of curriculum areas. Here, students produce their own multimedia texts in the form of websites or CD-ROMs, often combining written text, visual images, simple animation, audio and video material. Vivi Lachs (2000), for example, describes a range of production activities undertaken with primary school students in learning about science, geography or history. These projects generally involve children "re-presenting" their learning for an audience of younger children in the form of multimedia teaching materials or websites. Yet although the children's productions frequently draw on elements of popular culture (such as computer games), the content of the productions is primarily factual and informational—resulting in a form of "edutainment."

Other potential uses of digital media have emerged from arts education. These projects often involve the participation of "digital artists" external to the school, and their primary emphasis is on the use of the media for self-expression and creative exploration. The implicit model here is that of the avant-garde multimedia art work, although (here again) students tend to "import" elements of popular culture. Rebecca Sinker (1999), for example, describes an online multimedia project which set out to develop links between an infant school and its community. The project was intended to mark the school's centenary, and to offer the children opportunities "to investigate their own families, community, histories and experiences, exploring changes and celebrating diversity." Using multimedia authoring software, the project brought together photography, video, drawing, story-telling, digital imaging, sound and text. Perhaps most significantly, the results of the project (in the form of a website) were available to a much wider audience than would normally have been the case with children's work.

These approaches are certainly interesting and productive, but there are two factors that distinguish them from the use of digital production in the context of media education. Firstly, media education is generally characterized by an explicit focus on popular culture—or at least on engaging with students' everyday experiences of digital media rather than attempting to impose an alien "artistic" or "educational" practice. In the case of the internet, this means recognizing that most young people's uses of the medium are not primarily "educational," at least in the narrow sense. Teachers need to recognize that young people's uses of the internet are intimately connected with their other media enthusiasms—and that this is bound to be reflected in the texts they produce.

Secondly, there is the element of theoretical reflection—the dynamic relationship between making and critical understanding that is crucial to the development of "critical literacy." In the context of media education, the aim is not primarily to develop technical skills, or to promote "self-expression", but to encourage a more systematic understanding of how the media work and hence to promote more reflective ways of using them. In this latter respect, media education directly challenges the instrumental use of media production as a transparent or neutral "teaching aid." In fact, these digital tools can enable students to *conceptualize* the activity of production in much more powerful ways than was possible with analogue media. For example, when it comes to video production, digital technology can make overt and visible some key aspects of the production process that often remain "locked away" when using analogue technologies. This is particularly apparent at the point of editing, where complex questions about the selection, manipulation and combination of images (and, in the case of video, of sounds) can be addressed in a much more accessible way. In the process, the boundaries between critical analysis and practical production—or between "theory" and "practice"—are becoming increasingly blurred (see Burn & Durran, 2006).

Conclusion

The kinds of work I have referred to in this article are by no means new. On the contrary, they draw on an existing practice in schools that has a long history (see Buckingham, 2003). As in any other area of education, there is both good and bad practice in media education, and there is currently an alarming shortage of specialist trained media teachers. Nevertheless, it is clear that effective media education depends upon teachers recognizing and respecting the knowledge students already possess about these media—while also acknowledging that there are limitations to that knowledge, which teachers need to address.

I have argued here for an extension of media literacy principles to digital texts. This certainly entails some adaptation in how we think about media literacy—in its conceptual apparatus, and its methods of study (for example, in how we think about "audiences" or how we address the medium of games). Nevertheless, the media literacy model puts issues on the agenda that are typically ignored or marginalized in thinking about technology in education—and particularly in the school subject of ICT. Media literacy provides a means of connecting classroom uses of technology with the "techno-popular culture" that increasingly suffuses children's leisure time—and it does so in a critical rather than a celebratory way. It raises critical questions that most approaches to information technology in education fail to address and thereby moves decisively beyond a merely instrumental use of technology.

Ultimately, however, my argument here is much broader than simply a call for media education. The metaphor of literacy—while not without its problems—provides one means of imagining a more coherent, and ambitious, approach. The increasing convergence of contemporary media means that we need to be addressing the skills and competencies—the multiple literacies—that are required by the whole range of contemporary forms of communication. Rather than simply adding media or digital literacy to the curriculum menu or hiving off information and communication technology into a separate school subject, we need a much broader reconceptualization of what we mean by literacy in a world that is increasingly dominated by electronic media.

References

Barton, D. (1994). *Literacy: An introduction to the ecology of written language.* Oxford: Blackwell.

Bazalgette, C. (1988). 'They changed the picture in the middle of the fight': New kinds of literacy. In M. Meek, & C. Mills (Eds.), *Language and literacy in the primary school.* London: Falmer.

Beavis, C. (1998). Computer games, culture and curriculum. In I. Snyder (Ed.), *Page to screen: Taking literacy into the electronic era* (pp. 234–255). Sydney: Allen & Unwin.

British Film Institute. (2000). *Moving images in the classroom: A secondary teacher's guide to using film and television.* London: British Film Institute.

Bruce, C. (1997). *The seven faces of information literacy.* Adelaide: Auslib Press.

Buckingham, D. (1989). Television literacy: a critique. *Radical Philosophy, 51,* 12–25.

———. (1993a). *Changing literacies: Media education and modern culture.* London: Tufnell Press.

———. (1993b). *Children talking television: The making of television literacy.* London: Falmer.

———. (2003). *Media education: Literacy, learning and contemporary culture.* Cambridge: Polity.

Burbules, N.C., & Callister, T.A. (2000). *Watch IT: The risks and promises of information technologies for education.* Boulder, CO: Westview.

Burn, A. (2004). From *The Tempest* to *Tomb Raider*: Computer games in English, media and drama. *English, Drama, Media, 1(2),* 19–24.

Burn, A., & Durran, J. (2006). Digital anatomies: analysis as production in media education. In D. Buckingham & R. Willett (Eds.), *Digital generations: Children, young people and new media* (pp. 273–293). Mahwah, NJ: Lawrence Erlbaum.

———. (2007). *Media literacy in schools: Practice, production and progression.* London: Paul Chapman.

Carr, D., Buckingham, D., Burn, A. & Schott, G. (2006). *Computer games: Text, narrative and play.* Cambridge: Polity.

Cope, B. & Kalantzis, M. (Eds.). (2000*). Multiliteracies: literacy learning and the design of social futures.* London: Routledge.

Eco, U. (1979). Can television teach? *Screen Education 31,* 15–24.

Fabos, B. (2004). *Wrong turn on the information superhighway: Education and the commercialization of the internet.* New York: Teachers College Press.

Gilster, P. (1997). *Digital literacy.* New York: Wiley.

Goodson, I., & Mangan, J.M. (1996). Computer literacy as ideology. *British Journal of Sociology of Education, 17(1),* 65–79.

Jenkins, H. (2006). *Convergence culture.* New York: New York University Press.

Kress, G. (1997). *Before writing: Rethinking the paths to literacy.* London: Routledge.

Lachs, V. (2000). *Making multimedia in the classroom: A practical guide.* London: Routledge.

Livingstone, S. (2004). The challenge of changing audiences: Or, what is the audience researcher to do in the age of the internet? *European Journal of Communication, 19(1),* 75–86.

Livingstone, S., van Couvering, E., & Thumim, N. (2005). *Adult media literacy: A review of the research literature.* London: Ofcom.

Lorac, C., & Weiss, M. (1981). *Communication and social skills.* Exeter: Wheaton.

Mackey, M. (2002). *Literacies across media: Playing the text.* London: Routledge.

McDougall, J. (2006). *The media teacher's book.* London: Edward Arnold.

Messaris, P. (1994). *Visual 'literacy': Image, mind and reality.* Boulder, CO: Westview.

Moore, D., & Dwyer, F. (1994*). Visual literacy: A spectrum of visual learning.* Englewood Cliffs, NJ: Educational Technology Publications.

Oram, B., & Newman, J. (2006). *Teaching videogames.* London: British Film Institute.

Salen, K., & Zimmerman, E. (2003). *Rules of play: Game design fundamentals.* Cambridge, MA: The MIT Press.

Sinker, R. (1999). The Rosendale Odyssey: multimedia memoirs and digital journeys. In J. Sefton-Green (Ed.), *Young people, creativity and new technologies,* London: Routledge.

Spencer, M. (1986). Emergent literacies: A site for analysis. *Language Arts, 63(5),* 442–53.

Tyner, K. (1998). *Literacy in a digital world.* Mahwah, NJ: Erlbaum.

Warlick, D. (2005). *Raw materials for the mind: A teacher's guide to digital literacy.* (4th ed.). Raleigh, NC: The Landmark Project.

Williams, J., Clemens, S., Oleinikova, K. & Tarvin, K. (2003). *The skills for life survey: A national needs and impact survey of literacy, numeracy and ICT skills.* London: Department for Education and Skills.

CHAPTER FIVE

Digital Literacy Policies in the EU— Inclusive Partnership as the Final Stage of Governmentality?

LEENA RANTALA AND JUHA SUORANTA

Introduction

Digital literacies has an appealing image in the public sphere. In political, policy and, increasingly, in academic discourses digitalization represents itself as the latest achievement in the history of human progress and development, like a culminating stage in the long revolution of human empowerment and struggle against the forces of 'nature.' It is almost as if digitalization not only promises salvation from servitude and impoverishment but also brings a solution to the haunting problems of innovation and creativity in the global competition between nation states and state unions, and between the west and the rest.

Furthermore it is believed that digital literacy is a vital ingredient of the competence of individuals and increases the selling power of national educational markets. National governments all around the globe emphasize strong literacy as a more-important-than-ever skill in today's knowledge-based societies. It is further alleged that literacy provides a foundation for skills development and lifelong learning and can help all citizens participate in the nation's economic prosperity and improve their quality of life. Moreover, it is empha-sized that individuals' literacy skills create new immaterial markets, generate digitalized business and help to produce new commodities. These claims are supported by politically motivated studies, largely based on the premises of human capital theory, which aim to show that investment in literacy, and edu-cation in general, are more important to economic growth over the long run than investment in physical capital. Thus well-being enters into the equation. How, then, could anyone be critical of or have doubts about the benefits of digital literacies promulgated by digital literacy policies? Seemingly, there is no international organization, national government or state union in today's world that could afford to ignore digital literacy in its present and future policy development.

Of course, the above lines are exaggerations, but only slightly so, for, at the moment, these are among the most typical themes attaching to digital lit-eracy in public political debate and in diverse policy documents. But as always, there is more at play than meets the eye. Questions concerning digital literacies have almost nothing to do with information and communication technologies, with those electronic and invisible binary strings of ones and zeros, and almost nothing to do with the above-mentioned promises, but they certainly have to do with the technocratic apparatus of governing people in the world of global economic competition and teaching individual "networking" competences. In this chapter we approach digital literacies as part of the political terminology of the day as they connect to the European Union's policies on information technologies, lifelong learning and education.

Accordingly, for us questions concerning digital literacy as governmental policies are questions of political rationalities. Elements of Foucault's concep-tual framework are particularly pertinent here and will be addressed in the following section.

In what follows we analyze the European Union's (EU) elearning and dig-ital literacy policies as they have been promulgated to date. This is a provisional analysis because the EU's policies and conceptions of digital literacy have not yet crystallized but, rather, remain on the move. To try and get the best fix possible on the present and the future we wanted to augment the extant docu-ment base and our document-based analysis (Lankshear & Knobel, 2004, pp. 54–55) with information based on the personal assessments of EU experts of the current state and future prospects of digital literacy policies. We employed what we think of as "an email enhanced Delphi method." We emailed the members of the two EU expert groups, those of the eEurope Advisory Group and the Media Literacy Expert Group, but without notable success. From the total pool of around eighty members we received only five email replies to our questions: What are, in your assessment, the EU's major achievements so far in the area of digital literacies? What are your prospects for the future developments in the area of digital literacies in the EU? What are your hopes and fears? What is, in your opinion, the best EU policy document regarding digital literacies? And why?

Governmentality and the EU's Education Policies

In order to detach and objectify ourselves as (potentially critical) researchers from the persuading language of EU's policy documents we turned to Michel Foucault's theory of power and, especially, to his idea of governmentality, as many others who have analyzed EU and other education policy texts have done (cf. Olssen, 2006; Popkewitz et al., 2006; Simons, 2006; Simons & Masschelein, 2006; Tuschling & Engemann, 2006). Among Foucault's strengths is his insistence on connecting empirical "mole" work with his spirited desire to theorize empirical observations. We concur with Étienne Balibar's belief that Foucault's tendency to dialectical thinking, mixing theoretical and empirical domains, reflected his genuine and continuous struggle with Marx (see Lemke, 2000, p. 1). In the process of this struggle he developed his concept of governmentality, which serves as a central frame of reference in our analysis. According to Foucault the concept of governmentality refers to human beings in their various relations, "their links, their imbrication with those things that are wealth, resources, means of subsistence, the territory [and] fertility" as well as customs and habits as well as ways of acting and thinking (Foucault, 2000, pp. 208–209).

In a lecture on the subject of governmentality Foucault distinguishes three moments of the concept. First, governmentality is the "ensemble formed by the institutions, procedures, analyses, and reflections, the calculations and tactics that allow the exercise of this very specific albeit complex form of power, which has its target population as its principal form of knowledge, political economy, and as its essential technical means, apparatuses of security" (ibid., pp. 219–220). Second, governmentality refers to the tendency that has led over a long period of time toward preeminence over all other forms of power such as sovereignty and discipline. Third, the concept is linked to the processes through which the state of justice (of the Middle Ages) has transformed via the administrative state (of Modern times) into the state of governmentality, or as Foucault puts it, the administrative state "governmentalized" (ibid., p. 220).

Foucault's conception observes that up until the 18th century government was a term widely used in various public arenas and scientific texts ranging from politics and philosophy to religion, medicine and pedagogy. In addition to management by the state, as Lemke has pointed out, "government" also signified problems of self-control, guidance for the family and for children, management of the household, directing the soul, etc. For this reason, Foucault defines government as conduct, or, more precisely, as "the conduct of conduct" and thus as a term which ranges from "governing the self" to "governing others" (Lemke, 2000, pp. 2–3). Governmentality is thus a set of diverse combinations of sophisticated technologies of power executed by governmentalized states or state unions like the EU. As Lemke (2000, p. 7) has noted:

> by coupling forms of knowledge, strategies of power and technologies of self the concept of governmentality for the study of neo-liberal governmentality allows for a more comprehensive account of the current political and social transformations, since it makes visible the depth and breath of processes of domination and exploitation.

For our purposes the following aspects of governmentality, and questions based upon those aspects, are central: What are the rationalities and strategies used in the discourse of "digital literacy?" (i.e., governmentality as political knowledge); How is the role of economics seen in digital literacy policies? (i.e., governmentality as economics); What are the forms of digitally created technologies of the self, if any? (i.e., governmentality as a technology of the self).

In the current conjuncture the EU has projected into its rhetorical agenda an idea of digital literacy in order to attain what has for years been called a knowledge society or learning society. Are digital developments the latest developments in governmentality as we know it? This is a key question for us in this chapter. In what follows we also use the three questions formulated in the previous paragraph to frame our analysis of some of the main EU policy documents concerning digital literacy. We begin our inquiry with some brief forays into the concept of digital literacy itself and its various transformations as well as into the history of literacy policies in the EU.

Digital Literacies

Since the 1960s literacy has been one of the central issues of numerous governmental and international policies. Prior to that, indeed for centuries, literacy has featured in accounts of the Enlightenment and modernity as a key construction for governing, taming people, or making them docile. Literacy has long had a double meaning. On one hand it has been viewed as a prime "tamer" in the hands of rulers and the church. On the other hand, conversely, it has been seen as one of the cornerstones of individual and social emancipation. As Raymond Williams once put it: there is no way to teach people to read the Bible that does not also enable them to read the radical press or popular magazines for that matter (Williams, 2005, p. 134). In the administrative state's political agendas, literacy has been linked to economic prosperity and growth, and economic competition between the nation states. In the age of neoliberal globalization this interpretation and use of the concept of literacy have been intensified.

From another direction it is interesting to note that the concept of literacy and basic assumptions behind it have undergone considerable change during recent decades. Examples of the terminological change from basic literacy to new literacies can be seen in such terms and neologisms as media literacy, information literacy, digital literacy, technoliteracy, computer literacy, electronic literacy, network literacy—or even agricultural, dance, legal or workplace literacy (cf. Bawden, 2001). These changes in the very concept have followed socio-economic as well as cultural and technological changes in western countries. Literacy studies have likewise ridden the waves of socio-economic-technological change.

The ideal of literacy as a main vehicle for employability and social cohesion was born after WWII. UNESCO played a central role in promoting literacy as well as measuring illiteracy. The concept of functional literacy appeared at first in UNESCO papers in the 1950s and initially referred to basic literacy skills for all in order to function in a changing society. Since the 1960s, however, the meaning of "functional" has changed, becoming attached to economic efficiency as literacy was seen as a tool to prepare people not only for citizenship but also for productive work and consumerism (Levine, 1986, pp. 25–35).

Alongside the development of digital technologies and new media texts these functional or mechanical views on literacy have been challenged by many fresh literacy movements, research traditions and conceptualizations. Perhaps the most influential has been the move from a functional-autonomous to an ideological-sociocultural view of literacy and its meaning (Scribner and Cole, 1981; Street, 2004). Behind these changes was a so-called 'social turn' in human and social sciences in the early 1970s. The research interest was growing away from individuals and their private minds towards interaction between people and their social practices, and eventually between people and information and communication technologies. This was the time when the sociocultural approach to literacy and what was later known as new literacy studies were, perhaps, emerging for the first time (Lankshear, 1999).

The change is highlighted and juxtaposed by Street's (1984) notion of two different models of literacy, namely, the 'autonomous' and the 'ideological':

> The autonomous model construes literacy as existing independently of specific contexts of social practice; having autonomy from material enactments of language in such practices; and producing effects independently of contextual social factors. Accordingly, literacy is seen as independent of and impartial toward trends and struggles in everyday life—a "neutral" variable. (Lankshear, 1999; cf. Street, 1984)

In the autonomous model, language is distanced both from the learner and the teacher and treated "as a thing." Following Street's distinction, Dighe and Reddi (2006) claim that in the autonomous model:

> external rules and requirements are imposed and the significance of power relations and ideology in the use of language, ignored. In this model, language is conceptualized as a separate, reified set of 'neutral' competencies, autonomous of the social context. With regard to schooled literacy as well as of most adult literacy programmes, it is the autonomous model of literacy that has generally dominated curriculum and pedagogy.

On the other hand, the ideological model "rejects the notion of an essential literacy lying behind actual social practices involving texts. What literacy *is* consists in the forms textual engagement takes within specific material contexts of human practice" (Lankshear, 1999, no page; our emphasis). Literacy is thus seen as being inextricably and contextually linked to cultural, political and hegemonic power structures. In this sense it has been argued that because people's relationships with media in the digital age are necessarily tied to social and cultural contexts, it is important to get beyond individual, skills-based literacy learning and approach literacy as a sociocultural phenomenon (see also Buckingham, 2003; Livingstone, 2003). Accordingly, reading and writing are not only based on individual skills; literacy is rather an active relationship or a way of orienting to the social and cultural world. Furthermore, reading and writing do not happen in social isolation but, in some fundamental respect, are inherent attributes of social practices. Literacy is always a necessary, although not sufficient, part of social practice as examples like writing a conference paper or reading a comic book affirm (Gee, 2003, pp. 14–15; see also Lankshear & Snyder, 2000; Johnston & Webber, 2006). This situated nature of literacy is a key tenet shared by researchers working in the New Literacy Studies tradition.

Besides the internal changes in the research parameters of literacy studies, the ideological model or sociocultural perspective—especially the works of Paulo Freire (1972) and other Third World literacy studies authors from the 1960s—has posed an antithesis to the research tradition that investigates literacy from the standpoint of it being an individual or personal possession or competence. As Lankshear (1999) reminds us, Freire:

> explicitly denounced psychologistic-technicist reductions of literacy, insisting instead that 'Word' and 'World' are dialectically linked, and that education for liberation involved relating Word and World within transformative cultural praxis. Freire asserted the impossibility of literacy operating outside of social practice and, consequently, outside processes of creating and sustaining or re-creating social worlds. For Freire, the crucial issues concerned the kinds of social worlds humans create in and through their language-mediated practices, the interests promoted and subverted therein, and the historical option facing education of serving as either an instrument of liberation or of oppression. (no page)

Recent developments in the theory of literacy have in various ways emphasized the multiplicity of literacies. This idea has its theoretical grounds in the sociocultural tradition and its empirical base lies partly in the proliferation of digitalized information and communication channels in the past few decades. Alongside language, as James Paul Gee (2003, pp. 13–14) writes, these new information and communication systems involve many other visual symbols, such as images, graphs and diagrams, and the skills to use and interpret them. In addition, these "texts" are multimodal, that is, they mix words, images and other forms of information. Hence, multiple literacies are needed because there are different ways of reading and writing diverse multimodal texts. One extremely interesting current strand in the world of multiple literacies leads to the continuously developing digital technologies people use in collaboration. These social media and software, or Web 2.0 resources (e.g., webblogs and wikis) created in "the digital age" extend possibilities to engage in creative production of texts and share information (cf. Suoranta & Vadén, 2007). The creative and productive nature of these new "literacy machines" emphasizes writing over reading: literacy is productive and sometimes subversive practice, not only reflective interpretation and construction of meanings of texts (Kellner & Share, 2005). In addition, as Cynthia Lewis (2007) has reminded us, new literacies "allow writers (users; players) a good deal of leeway to be creative, perform identities, and choose affiliations within a set of parameters that can change through negotiation, play, and collaboration" (p. 231).

A Brief History of eLearning Policies in the EU

Before the eLearning initiative was developed in 2000, two action lines existed in the EU with respect to elearning: European funded research on new technologies and learning from the 1980s, and diverse alliances and co-operation in the field of education (i.e., the Erasmus program). A third factor in creating the EU eLearning initiative was the idea of lifelong education as a key European policy under the Lisbon Agenda (2000). This aimed at adapting European educational and training systems to a knowledge-based economy and society, with the understanding that "society's economic and social performance would be determined by the extent in which its citizens and its economic and social forces can use the potential of new technologies, how effectively they incorporate them into the economy and build up a knowledge-based society" (Díaz, 2006, p. 121).

Maruja Gutiérrez Díaz (2006) has advanced a useful analysis of the EU's eLearning policy focusing on three storylines. These are educational transformation, technological change and political co-operation, respectively. In the storyline of educational transformation, the first notable action is the eLearning initiative in 2000, with its four elements: (1) ICT infrastructure and equipment, (2) training at all levels, in particular teachers and trainers, (3) quality European contents and services and (4) European networking and co-operation. As a preparatory and exploratory action this initiative made three different project calls. Selected projects concentrated on higher education, media literacy and quality, observatories and networks. In addition, the eLearning initiative worked together with information and communication technology and media industries. The eLearning Industry Group included companies like IBM, Nokia and [Finnish Media Corporation] SanomaWSOY. The aim was to create partnerships with public and private stakeholders. Media literacy issues were dealt with as a part of this larger eLearning initiative. In the media literacy branch, the goal was to create a cultural, humanistic approach to the new digital culture, in contrast to a technological approach. Special attention was paid to new digital media and to integrating ICT and media literacy into school curricula.

Within this storyline of educational transformation, the eLearning Action Plan in 2001 was created with the aim of facilitating joint monitoring and co-operation of programs and instruments and better coherence and visibility of eLearning. There was also a need to construct a common understanding of the concept of eLearning. Furthermore, the eLearning program 2003 was generated along four lines of action: (1) school twinning via the internet (45% of the budget), (2) virtual campuses (30%), (3) promotion of digital literacy (10%) and (4) transversal actions, i.e., observatories, EFQUEL—the European Foundation for Quality in eLearning—with 7.5% of the budget. Uzunboylu (2006, no page) has summarized these e-learning decisions as follows:

> E-learning in Europe has focused on instituting practices that benefit schools and public services. European Councils are seeking to use ICT and the WWW strategically, not merely as means for everyday use. The e-Learning Action Plan and the e-Learning Program have been used to integrate ICT for education and training in European countries. The use of these strategies suggests that e-learning yields positive results. The EC has assumed an important role in planning, designing, implementing, and evaluating e-learning and in financially supporting its widespread implementation. The e-Learning Action Plan plays an important role in guiding European e-learning for achieving established goals and provides an important resource for member states. This plan also enables the exchange of knowledge and experiences related to key factors in using ICT for education and training, including financing infrastructures, purchasing equipment, providing net-work access, training strategies, supporting the development of instructional content and services, evaluating teaching methodology, and advancing further research.

The second eLearning storyline, dealing with eLearning in the wider context of technological change, is developed in the eEurope 2002 action plan. Two important ideals in the plan from the standpoint of education were to get "European youth to the digital Era" and to ensure a "faster internet for researchers and students." To achieve these goals technological and regulatory objectives were established with respect to considerations like the number of pupils per PC in a school. The action plan produced two Eurobarometer flash surveys for schools (in 2001 and 2003). A subsequent technological change-related program, eEurope 2005, had three subthemes: eGovernment, eHealth, and eLearning. It produced the third schools' Eurobarometer and held the 2005 eLearning Conference.

The third eLearning storyline, European political co-operation, plays out within the Education and Training 2010 program. It deals with common concerns and priorities inside the EU: quality, accessibility, and connecting to society. The ICT group launched under the initiative seeks a common understanding of ICT policy and practice and suggests four issues to improve education systems: (1) embed ICT policies and strategies into long term educational objectives; (2) ensure new support service for education; (3) empower educational actors and train them for the management of change; (4) develop research, establish new indicators and provide access to results. Furthermore, the ICT Cluster inside Education and Training 2010 concentrates on peer learning activities. Digital competence is one of the key competencies identified in the European framework for key competencies for lifelong learning.

Future perspectives for eLearning in the EU include plans for the integrated lifelong learning program 2007–2013. This will aim at general mainstreaming of ICT projects within the sectoral programs and at providing support for innovations. Beyond that, according to Díaz (2006, p. 149), there will be a shift from technological to cultural issues. This includes an aim to understand lifelong learning more as a culture than as a matter of instrumental training, since "it is no more a matter of why but of what and how." Although the hype of eLearning seems to be waning, faith in the potential of ICTs in learning has remained intact, notwithstanding acknowledgement of digital divides in different parts and among different populations of Europe.

As is evident in this brief history of eLearning policies in the EU, European policies are typified by the double strategy of combining social and economic dimensions. Rodrigues (2006, pp. 412–413) describes this double strategy in the following way:

> This model is the outcome of a long and complex historical process trying to combine social justice with high economic performance. This means that the social dimension should be shaped with the purpose of social justice, but also with the purpose of contributing to growth and competitiveness. Conversely, growth and competitiveness are crucial to support the social dimension and should also be shaped to support it. This also means that there are different choices in both economic and social policies which evolve over time and must be permanently under discussion, political debate and social dialogue.

Consistent with this, Manuel Castells and Pekka Himanen (2002) have maintained that the European combination of social and economic dimensions, and especially the Nordic welfare model, can be a sustainable ground for the development of a knowledge-based society if technological innovations are given a chance to flourish. And, as they claim, the future welfare model needs to be based on a sufficiently competitive and innovative knowledge-based economy in the global markets.

Reading the Contents of Digital Literacy Policies: From a Basic Skill to a Key Competence for Lifelong Learning

In what follows we provide a thematic reading of the EU's digital literacy rhetoric. We use numerous quotations from original sources to render the style, tone and, so far as possible, the substance of the given policy documents with respect to digital literacy. (Readers should not be surprised or feel guilty if they become a little bored reading the policy content, since this was our own experience during the process of preparing for this part of the chapter.)

Every Citizen of the Learning Economy Must Be Digitally Literate

To date there is no general paper or agreement on the policy, or substantial dimensions, of digital literacy inside the EU. But as we mentioned above, the process of defining the concept of digital literacy is currently in the hands of an expert group and their work is still in progress. As soon as they complete their work, the administration will commence its own work in molding policy papers for the digital literacy policies in the Union.

It is, however, worth noting that as early as 2000 the Lisbon Strategy, or Lisbon Agenda, acknowledged digital literacy as a concept and core component of future policy initiatives in Europe. The strategy was adopted for a ten-year period in Lisbon, Portugal, by the European Council. The strategy has three "pillars". First, there is an economic pillar concerned with preparing "the ground for the transition to a competitive, dynamic, knowledge-based economy. Emphasis is placed on the need to adapt constantly to changes in the information society and to boost research and development." Second, there is a social pillar concerned with "[modernizing] the European social model by investing in human resources and combating social exclusion. The Member States are expected to invest in education and training, and to conduct an active policy for employment, making it easier to move to a knowledge economy." (Presidency Conclusions on the Lisbon Strategy by Theme, 2000–2004, p. 22) Finally, an environmental pillar was added at the Gothenburg European Council meeting in June 2001, which draws attention to ecologically balanced and sustainable development, or to "the fact that economic growth must be decoupled from the use of natural resources." (ibid.)

The Lisbon Strategy leaned strongly to the idea of a "learning economy" as a version of the more familiar "knowledge economy," and it broadly aimed to "make Europe, by 2010, the most competitive and the most dynamic knowledge-based economy in the world." In its economic accent it was maintained in the Strategy that the shift to "a knowledge-based economy is of crucial importance for competitiveness and growth and for building a more inclusive society" (ibid., p. 22). Furthermore, the Strategy emphasized that "the success of the knowledge society also depends on high levels of digital literacy and on creating conditions in areas such as network security and data protection and privacy, in which people have confidence in using new services" (p. 22). Several practical recommendations were advanced, among them new basic skills: "a European framework should define the new basic skills to be provided through lifelong learning: IT skills, foreign languages, technological culture, entrepreneurship and social skills" and "a European diploma for basic IT skills, with decentralised certification procedures, should be established in order to promote digital literacy throughout the Union" (p. 87). In addition it was important to learn "at least two foreign languages from an early age," and "establish a linguistic competence indicator in 2003" and support development of digital literacy and generalization of an internet and computer user's certificate for secondary school pupils (p. 94).

Another resolution from the same year as the Lisbon Strategy, the "eLearning: designing tomorrow's education" statement, followed and further specified the Lisbon Strategy. It sought "to mobilize the educational and cultural communities, as well as the economic and social players in Europe, in order to speed up changes in the education and training systems for Europe's move to a knowledge-based society." The first stage in this move was to promote acquisition of the confident use of the internet and other new tools for accessing knowledge and, in addition, the widespread development of a 'digital literacy' which was to be "adapted to the different learning contexts and target groups." The initiative compared the industrial societies in terms of ensuring that all citizens had adequate conventional and digital literacy. According to the eLearning Initiative (pp. 3–4), all citizens need to be "properly versed in the three Rs" and:

> the emergence of the knowledge-based society implies that every citizen must be 'digitally literate' and [possess] basic skills in order to be on a better footing in terms of

equal opportunities in a world in which digital functions are proliferating. This is high on the list of priorities if we are to enhance cohesion and employability in our societies as opposed to creating fresh divisions.

Stepping Up the Training Drive at All Levels

The high costs of telecommunications were noted in the eLearning Initiative as an obstacle to the use of the internet and the spread of digital literacy. Several objectives for adapting education and training systems to the knowledge-based society were set. These included training "a sufficient number of teachers in the use of internet and multimedia resources," ensuring "that schools and training centres become local centres for acquiring knowledge which is versatile and accessible to everyone, using the most appropriate methods tailored to the broad diversity of the target groups"; adopting "a European framework to define the new basic skills which lifelong learning must make it possible to acquire" (e.g., information technologies, foreign languages, technical knowledge); defining "ways of encouraging mobility among students, teachers, trainers and researchers, through the optimal use of Community programmes, by removing obstacles and by increased transparency for the recognition of qualifications and periods of study and training"; preventing "the gap from constantly widening between those who have access to new knowledge and those who do not, by defining priority actions for certain target groups (minorities, the elderly, the disabled, the under-qualified) and women"; and providing pupils "with broad digital literacy by the end of the year 2003" (p. 7).

The *eLearning Action Plan: Designing Tomorrow's Education* was a communication from the Commission to the Council and the European Parliament (March 2001). It aimed at stepping up "the training drive at all levels, especially by promoting universal digital literacy and the general availability of appropriate training for teachers and trainers, including technology training as well as courses on the educational use of technology and management of change" (p. 3). It emphasized new technical, intellectual and social skills that "are becoming essential for living, working and participating actively in a knowledge society." It noted that while the scope of these new skills reaches "well beyond 'digital literacy,' they are the basis on which it depends. They fall into the broader category of 'new basic skills' (foreign languages, entrepreneurship, etc. as above) to be acquired in a process of lifelong learning. Discriminating and responsible use of the new technologies constitutes one of these new basic skills" (p. 11).

A Council Resolution of 13 July, 2001, requested an interim report (Commission Staff Working Paper). "E-Learning—Designing Tomorrow's Education" was presented in 2001 and affirmed that "the provision of e-learning to all citizens and ensuring digital literacy for every worker is part of the objectives of the European Employment Strategy" (p. 8).

The mid-term follow-up report of 2003 announced (p. 2) that in:

> launching the initiative 'eLearning: Designing Tomorrow's Education', with its corresponding Action plan for 2001–2004, the Commission laid the foundations for concrete and sustainable action, through a set of specific measures. In proposing the eLearning Programme 2004–2006, the Commission aims to strengthen this work by focusing attention on Digital literacy, School twinning, and Virtual campuses, whilst reinforcing its monitoring of the eLearning Action Plan.

The follow-up report also affirmed that "The eEurope 2005 Action Plan was launched to continue the work of eEurope, promoting the use of broadband communications and services, in order to improve the effectiveness and efficiency of public services." It was supposed to direct effort toward the three policy priorities of eLearning, eGovernment and eHealth. The plan defined actions to support the re-skilling of the workforce using e-learning and the deployment of virtual campuses and was planned to be an instrument in enhancing digital literacy and building virtual campuses (p. 5). It concluded that progress has been made, but the real work was only about to start:

> The eLearning Initiative has launched a number of activities to support the work under the eLearning Action Plan and the recent evaluation of the first projects has highlighted their positive contribution. (...) E-learning is starting to become mainstream in our education and training systems. Connectivity and equipment are no longer the central issues, as our focus moves to pedagogy, content, quality assurance and standards, teacher/trainer training and continuous development, organisational change and the transformation of education and training processes. Much progress has been made and yet many would admit that the real work is only now beginning. E-learning is coming of age and we are moving from preparation to practice; from e-learning pilots to enhanced, sustainable education and training programmes (pp. 12–13).

Lifelong Learning for Specific Target Groups: Emphases on Economy and Education

The objectives of the multi-annual E-Learning program for 2004–2006 were "to identify the actors concerned and inform them of ways and means of using e-learning for promoting digital literacy and thereby contribute to strengthening social cohesion and personal development and fostering intercultural dialogue" (Decision No 2318/2003/EC of the European Parliament and of the Council, December 2003). It was further maintained that digital literacy actions in this area "will address the contribution of ICT in school and more broadly in a lifelong learning context, in particular for those who, owing to their geographical location, social situation or special needs, do not have easy access to those technologies" (ibid.). The Annex to the document stated that:

> Action in this field must cover both conceptual and practical issues, from the understanding of digital literacy to identification of remedial actions for specific target groups. Digital literacy is one of the essential skills and competences needed to take an active part in the knowledge society and the new media culture. Digital literacy also relates to media literacy and social competence, as they have in common objectives such as active citizenship and the responsible use of ICTs. (no page)

The eEurope 2002 Action Plan (2000) addressed the challenge of achieving full employment through "a radical transformation of the economy and skills to match the opportunities of the new economy." This action plan took education and training as primary means to reach these goals. At the same time it was stressed that because the results of education could be only realized in the longer term, in order to have faster changes, something more had to be done. Hence, jobs for information technology professionals were taken into the agenda, because "studies on the skills gap indicate that Europe currently has around 800,000 vacancies, expected to grow to around 1.7 million by 2003 unless action is taken." Besides the demand for information technology professionals, actions were also targeted toward citizens, for "digital literacy is an essential element of the adaptability of the workforce and the employability of all citizens." The responsibility of this digital literacy training was put in the hands of enterprises, thus enhancing life-long learning in work places. The enterprises were suggested to be promoted with promises for awards for companies that are "particularly successful in developing human resources" (p. 16).

The policy measures introduced in the eEurope 2005 Action Plan suggested that responses to the eEurope 2002 targets had been positive with respect to the Trans-European networks that aimed to connect national research and education networks. However, the work to provide access to the internet and multimedia resources for schools, teachers and students appeared to be only at a beginning, despite the goal set by the Barcelona European Council to ensure full access by end of 2003 (comprising a ratio of 15 pupils per on-line computer in every school inside the EU). Digital literacy was mentioned in the footnote of this action plan (p. 11) as follows:

The Barcelona European Council also requested to develop digital literacy through the generalization of an internet and computer user's certificate for secondary school pupils and to undertake a feasibility study to identify options for helping secondary schools to establish or enhance an internet twinning link with a partner school elsewhere in Europe.

Proposed actions in the eEurope 2005 action plan included "re-skilling for the knowledge society." This referred in particular to launching actions designed to provide adults (notably, the unemployed and women returning to the labor market) with key skills (namely, basic computer skills, or digital literacy) and higher-order skills like teamwork, problem solving, project management), and to improve their employability and overall quality of life. The document underlined the idea that these actions would take advantage of the possibilities offered by e-learning (p. 12).

Searching for Consensus Among EU Countries: Focus on Teachers and Teacher Education

"Education and Training of Teachers and Trainers" is part of the Education and Training 2010 program, which aimed to set common educational standards for European schooling systems. This goal has been justified partly by political rhetoric and partly by surveys from different EU countries, which indicate that there are discrepancies between education policies and national schooling systems (i.e., diversity of student intake in e-learning and differences in the teaching environments). Thus recommendations from the EU call for more attention to legal requirements, changing dimensions of learning, and new competencies in each EU country. Particular focus has been given to teacher's work beyond the classroom (e.g., curriculum and organization development and co-operation with social partners) as well as to learning outcomes. The two suggested agendas for the participating countries based on the work of the Expert Group were: (1) "Identifying the skills that teachers and trainers should have, given their changing roles in the knowledge society" and (2) "Providing the conditions which adequately support teachers and trainers as they respond to the challenges of the knowledge society, including through initial and in-service training in the perspective of lifelong learning." A number of specified teacher and trainer competencies were already defined: motivation to learn beyond compulsory education; learning how to learn/learning in an independent way; information processing (with a critical eye); digital literacy;

creativity and innovation; problem-solving; entrepreneurship; working with others; communication and visual culture. (Working Group's Report 2003, 8, 18–19, 40, 49).

The theme of information and communication technologies is also addressed in the Education and Training 2010 program. This concludes that the reports from different EU countries related to ICT policies indicated that there were common trends across them all. Such trends included addressing educational issues considered strategic for the country, like targeting teacher education as a key focus for developing integration of ICT in education. However, it was also recognized that no consensus existed on how digital literacy should be generally defined and addressed in the school curriculum, although training students in ICT basic skills is regarded as one of the objectives for students to "enter the digital and media culture." Some policies are seen as focusing more on computer literacy, while others extend education to all media. It is also stated that:

> Digital literacy is increasingly defined in terms of intellectual capacities and not just in terms of physical access. In the same manner, the digital divide is increasingly related to the equity access to information in the educational, scientific, economic, social, political and cultural fields. Obviously, 'accessing' to information does not mean 'mastering' related knowledge, but access already appears as being a political goal for many countries or regions of Europe. (Working Group's Report 2003, p. 8, 20–21)

Tuschling and Engemann (2006, p. 452) state that the overall paradox in the EU's governmentality is built around the double ideology of individualism and totalization; that is, at the same time as people are supposed to enjoy their new freedoms and responsibilities they are confronted with an ever-growing field of individual incentives and competition which maximize their own 'life-chances' and minimizes their costs to the state. There are further paradoxes in the EU's education policies. The key concept seems to be informal learning; learning can and will occur every day in every way. It is an anthropological fact. This does not mean, however, that the EU would try to set learning free as was the case with such learning society protagonists in the late 1960s as Rudi Dutschke. The purpose of the EU's education policy is to administer informal learning by setting its institutional premises. Three issues are involved here: "changing the field of learning in order to totalize learning to all imaginable situations," "initiating a change in the self-performance of individuals" so that they are able to act in the newly totalized learning situations, and inventing inter-institutionalizing techniques "that allow both individuals and institutions to inscribe, store, process and transfer actions as learning" (ibid., p. 460).

A Key Competence for Using Information Society Technology

Naming the key competencies of the knowledge society is likewise on the agenda of the Education and Training 2010 program. Neither the Recommendation of the European Parliament nor of the Council (2006) on key competences for lifelong learning nor the Commission's proposal for a Recommendation on Key Competences for Lifelong Learning (November 2005) mentions the term digital literacy. However, literacies are addressed as follows:

> Learning to learn skills require firstly the acquisition of the fundamental basic skills such as literacy, numeracy and ICT skills that are necessary for further learning. Building on these skills, an individual should be able to access, gain, process and assimilate new knowledge and skills. This requires effective management of one's learning, career and work patterns, and, in particular, the ability to persevere with learning, to concentrate for extended periods and to reflect critically on the purposes and aims of learning. Individuals should be able to dedicate time to learning autonomously and with self-discipline, but also to work collaboratively as part of the learning process, draw the benefits from a heterogeneous group, and to share what they have learnt. Individuals should be able to organize their own learning, evaluate their own work, and to seek advice, information and support when appropriate. (Commission proposal for a Recommendation on Key Competences for Lifelong Learning, 2005, p. 17)

It is emphasized that when it comes to the definition of the framework for key competencies in broader terms it is impossible and irrelevant to distinguish between basic and advanced levels of mastery of a competence. This is because:

> the term 'basic' refers to something that depends on the requirements of the situation and circumstances: mastering a skill well enough to solve a problem in one situation might not be enough in another situation. In a constantly changing society, the demands faced by an individual vary from one situation to another and from time to time. Therefore, in addition to possessing the specific basic skills for accomplishing a certain task, *more flexible*, *generic* and *transferable* competences are needed to provide the individual with a combination of skills, knowledge and attitudes that are appropriate to particular situations. (Key Competences for Lifelong Learning. A European Reference Framework. November 2004, p. 4)

Digital literacy is seen to be a good example of the situational nature of key competences because there are only relatively few situations where basic ICT skills are sufficient. In most cases the effective use of ICT requires an appropriate level of critical thinking and a wider understanding of media.

Finally, the eight key competencies proposed in the Recommendation of the European Parliament and of the Council (2006) comprise: (1) communication in the mother tongue, (2) communication in foreign languages, (3) mathematical competence and basic competences in science and technology, (4) digital competence, (5) learning to learn, (6) social and civic competences, (7) a sense of initiative and entrepreneurship, and (8) cultural awareness and expression. As it is defined in the document, digital competence can be read as a proxy for digital literacy:

> Digital competence involves the confident and critical use of Information Society Technology (IST) for work, leisure and communication. It is underpinned by basic skills in ICT: the use of computers to retrieve, assess, store, produce, present and exchange information, and to communicate and participate in collaborative networks via the Internet. (ibid., p. 6)

In addition, essential knowledge, skills and attitudes related to digital competence are to consist in "understanding and knowledge of the nature, role and opportunities of IST in everyday contexts: in personal and social life as well as at work" including "main computer applications such as word processing, spreadsheets, databases, information storage and management, and an understanding of the opportunities and potential risks of the internet and communication via electronic media (email, network tools) for work, leisure, information sharing and collaborative networking, learning and research." Moreover, individuals should "understand how IST can support creativity and innovation, and be aware of issues around the validity and reliability of information available and of the legal and ethical principles involved in the interactive use of IST." (ibid.) The skills needed are seen to include "the ability to search, collect and process information and use it in a critical and systematic way, assessing relevance and distinguishing the real from the virtual while recognizing the links." Furthermore, individuals should "have skills to use tools to produce, present and understand complex information and the ability to access, search and use internet-based services" and "be able to use IST to support critical thinking, creativity, and innovation." It is also stated that "use of IST requires a critical and reflective attitude towards available information and a responsible use of the interactive media. An interest in engaging in communities and networks for cultural, social and/or professional purposes also supports this competence." (ibid.)

EU Digital Literacy Policy as 'Inclusive Liberalism'

Access issues, infrastructure and resources seem particularly to be emphasized among the earlier e-learning initiatives. This reflects a technological determinism or, at least, a deep faith in the power of new technologies to change education and learning for the good of Europe's economic and social welfare. The focus of calls for action has been in school curricula and teacher training, although "the special groups" including minorities, women, the elderly, the disabled, etc., are quite often mentioned. However, as a whole, digital literacy concerns all Europeans labeled as citizens or (future) work force. The other central theme in the documents appears to be defining the concepts: basic skills in the earlier papers and, later, competencies for the knowledge society. All the themes of democratic participation and active citizenship, knowledge economy, competition and individual choices, and life-long learning, cultural self-expression and personal fulfillment (Livingstone et al., 2008) appear to go hand in hand in the EU documents. Hence, it seems as if all the spheres of human life from civil society to work and leisure time were covered and *governed* in the documentation on e-Europe and digital literacy. In this respect, policy documentation is fundamentally not about information and communication technology *as such*. Rather it is really about "information society technologies" in the service of the economy-technology complex.

The EU is already a huge paper and document mill with regard to digital literacy proposals and initiatives after only few years of policy development directed at digitalization. Moreover, it is shiningly clear that EU has harnessed the concept of digital literacy to the vehicles of global economic competition between Europe and others. This is the recurring theme in the documents analyzed. The idea is that by applying ICT and developing digital literacy as an individual skill and competence it is possible in the long run to train an efficient and competitive work force able to meet the needs of the global markets. Based on our readings, the view of literacy in the EU really is autonomous, functional, individual and competence-based, with few signs of more socially or ideologically oriented views. Thus, the idea is that introducing literacy to illiterate people enhances their cognitive skills, improves their economic prospects, and makes them better citizens, regardless of the social and economic conditions that accounted for their illiteracy in the first place (see Street, 2003, pp. 77–78). But from a critical point of view it is erroneous to suggest that literacy can be given neutrally—if it can be "given" at all; maybe it can only be achieved, or more precisely, acquired with others—and its social effects only subsequently experienced. In any case, this strand of digital literacy debate pertains to what can be labeled as "prolonged exchange value of well-educated citizens." This is the overriding discourse of digital literacy in the present EU digital literacy policies, which can be argued to represent, instead of 'hard way' neo-liberalism, a sort of 'inclusive' liberalism (Porter & Craig, 2004). Inclusion captures something essential of the working of governmentality inscribed in the EU documents. Three related inclusions seem to be most important from the point of view of digital literacy policies.

The first inclusion in the EU digital literacy documents incorporates the poor within 'the global economy.' For nation states this means "adopting world trade rules and conservative fiscal policies, removing trade barriers and opening capital markets, but not necessarily removing migration or trade barriers in the core." From the standpoint of individuals, this primary inclusion "is inclusion in labour markets, or in training for these, a preparation which now begins in the social investments made all the way from (before) the cradle, to the community to the (global) workplace and economy." (Porter & Craig, 2004, p. 4)

The second inclusion is ideological and political. It reaches

> well beyond mere market liberalisation, to include concerns about security, stability, risk, safety, inclusion and participation. All of these are de-politicized, consensual rationales, absolutely suited to a global liberal order without serious ideological rival. Here, the ideological and political task is to imagine and create ways to offer the most excluded of the poor some stake in the wider liberal order, while at the same time protecting it from legitimate contest. Great efforts are made to be seen to 'include' those classic liberal subjects, the vulnerable: the excluded, the poor, the marginal, the child. Whereas a previous neo-liberalism would have left these to sink or swim in the free market, 'inclusive' liberalism won't let them get away so easily. Their right to be included comes with obligation. (ibid., p. 4)

Among these obligations, from the individual's point of view, are entrepreneurship, lifelong learning and up-to-date training for individual competitiveness. The third form of inclusion is practical and governmental. It involves:

> the active reconfiguring of structures of society and governance along more global, 'inclusive' liberal lines. . . . This re-structuring is often achieved through 'information rich' technical programmes and measures (e.g., Participatory Poverty Assessments, individualised work tests). Individuals and their local communities are then reintroduced to wider governance and market relations in subordinated, disaggregated ways, as localities with their own strategic plans, as regions coordinating their own service delivery, as partners in social governance or community development. Again, this integration is not simply neo-liberal market integration: it is the active constitution of rules, relations, and domains, which are imagined to be social as well as economic, and within which you are 'included.' (ibid., p. 4)

Porter and Craig add that this imagined and governmentally manufactured inclusion "obscures real social differences and conflict and collapses local authority into technically bounded domains." For those who remain outside the inclusive state (union) liberalism, or for some reason fail to participate or do not want to participate, "there is the obverse face of 'inclusion': entrapment and labeling if not as 'terrorist,' then at least deserving moral sanction and policing, and suspension of even meagre charity and benefits." (Porter & Craig, 2004) In 'inclusive' liberalism the governing state is replaced with the partnering state, which "while maintaining close affinities with markets, evinces other concerns as well, which it represents as being compatible with, or even essential for successful market development. These concerns include the security, care, and upbringing of the future knowledge economy workers" (Roelvink & Craig, 2004, p. 4).

There is a sub-theme subsumed within the primary, economically determined one. This is the theme of digital literacy as an extension of a humanism renaissance. This point of view has been highlighted by José Manuel Pérez Tornero (2004, pp. 57–58), who maintains that at present "we have the opportunity to restore and move our entire previous cultural heritage to the digital world." He claims that Europe is facing a new kind of humanism in which digital literacy and digital culture will support a humanistic understanding of culture. Information technologies are seen not as technical tools but as expansions of humanism's cultural heritage. In this register, Tornero (ibid.) further claims that in the context of media education digital literacy presents "the opportunity to assert a new identity and a new civic responsibility; that is, a new statute for the individual and for their rights and obligations." New information technologies can be means to enhance human rights and peoples' citizenship by "expanding their knowledge, increasing their freedom and allowing mutual recognition."

On the other hand, however, these technologies can also increase surveillance and monitoring of ordinary life. Hence, Tornero observes that it is vital to guide the development and implementation of digital literacy policies by affirming democratic rights and ways to control the controllers.

> Accordingly, digital literacy has to enable us rethink social relations, duties and rights and pave the way for learning new values; values that are more solid and steadfast in their equity and solidarity, respectful of human dignity. With respect to this social dimension of digital literacy, the incorporation of ICT in institutions and society must provide opportunities not only to increase efficiency and accelerate certain existing processes, but also to rethink such processes and change them, adapting them to human and social needs. (ibid.)

Digital literacies as social practices are constituted and enacted within a historical and structural context shaped by the mode of production and class relations, which change over time. These phenomena should be analyzed in the global context for they have global impacts. Different classes and groups have different interests in a digital world, and these are often contradictory and in conflict. The conflicts in a digital world are reflected at the state level as well as the global level. Hence, national and regional public policies (such as the EU policies) should be analyzed in terms of the various inequalities they directly or indirectly produce. Intellectual and cultural life is formed by the capitalist mode of production, and the struggle for ideological hegemony plays out in both the material world and the world of ICTs, as well as at the levels of globality, the state, and civil society. At the latter level many organizations of civil society "seek to transform people's understanding of society and thereby engage their support in struggles to change society" (Youngman, 2000, p. 30). The message we take from all this is that the ideological game concerning digital literacy is not over. It is only starting.

Concluding Remark

This leads to our concluding remark: is there any point in emphasizing the idea of multiple literacies in the EU policy documents and decision-making other than at the merely rhetorical level, if the reality is defined by the tyranny of the market? How, if at all, can we act to ensure diverse literacies, local literacies, ethnic literacies and other form of literacies that do not necessarily link directly to the instrumental ends of an efficient work force, economic values or new modes of immaterial production but, rather, fulfill other goals in the lives of millions? The least we can do as educators and researchers in the field of new media machines is to extend our research. As Hiiseyin Uzunboylu (2006) puts it, this is a call "to determine technological, pedagogical, social-economical, and cultural affects of e-learning throughout countries in the EU." Furthermore, we concur with Norman Fairclough's (2005) belief that it is vitally important "to go beyond public policy documents, and to research the operationalization of discourses such as the 'information society' and the 'knowledge economy' not only by examining government initiatives such as the 'e-government' website but also, crucially, through ethnographic research which can give insights into the relationship between discourses, rhetoric, and reality." Thus, perhaps, we could begin by asking: Whose literacy, whose digitalization?

Original Documents Referred to in This Chapter

Extracts from Presidency Conclusions on the Lisbon Strategy by Theme. European Councils: Lisbon (March 2000) to Brussels (June 2004). Available at <http://ec.europa.eu/growthandjobs/pdf/thematic_lisbon_conclusions_0604_en.pdf> (Retrieved October 2, 2007)

eLearning—Designing tomorrow´s education. Communication from the Commission. Brussels 24.5.2000 COM (2000) 318 final. Commission of the European communities. Available at <http://europa.eu/eur-lex/en/com/cnc/2000/com2000_0318en01.pdf> (Retrieved October 2, 2007)

The eLearning Action Plan. Designing Tomorrow´s Education. Communication from the Commission to the Council and the European Parliament. Brussels, 28.3.2001 COM (2001) 172 final. Commission of the European communities. Available at http://eurlex.europa.eu/LexUriServ/site/en/com/2001/com2001_0172en01.pdf> (Retrieved October 2, 2007)

eLearning: Designing Tomorrow´s Education. An Interim Report. As requested by the Council Resolution of 13 July 2001. Commission Staff Working Paper. Brussels, 28.2.2002 SEC (2001) 236. Commission of the European communities. Available at <http://ec.europa.eu/education/programmes/elearning/sec_2002_236_en.pdf> (Retrieved October 2, 2007)

eLearning: Designing Tomorrow´s Education. A Mid-Term Report. As requested by the Council Resolution of 13 July 2001. Commission Staff Working Paper. Brussels, 30.7.2003 SEC (2003) 905. Commission of the European communities. Available at <http://ec.europa.eu/education/programmes/elearning/doc/mid_term_report_en.pdf> (Retrieved October 2, 2007)

Decision No 2318/2003/EC of the European Parliament and of the Council of 5 December 2003 adopting a multiannual programme (2004 to 2006) for the effective integration of information and communication technologies (ICT) in education and training systems in Europe (eLearning Programme). Official Journal of the European Union 31.12.2003. Available at <http://europa.eu/eur-lex/pri/en/oj/dat/2003/l_345/l_34520031231en00090016.pdf> (Retrieved October 2, 2007)

eEurope 2002. An Information Society for All. Action Plan prepared by the Council and the European Commission for the Feira European Council. 19–20 June 2000. Brussels, 14.6.2000. Council of the European Union. Commission of the European Communities. Available at <http://ec.europa.eu/information_society/eeurope/2002/action_plan/pdf/actionplan_en.pdf> (Retrieved October 2, 2007)

eEurope 2005: An information society for all. An action plan to be presented in view of the Sevilla European Council, 21/22 June 2002. Communication from the Commission to the Council, the European Parliament, the Economic and Social Committee and the Committee of the Regions. Brussels, 28.5.2002 COM 82002) 263 final. Commission of the European Communities. Available at <http://ec.europa.eu/information_society/eeurope/2002/news_library/documents/eeurope2005/eeurope2005_en.pdf> (Retrieved October 2, 2007)

Education and Training 2010. European Commission. Website of the program available at <http://ec.europa.eu/education/policies/2010/et_2010_en.html> (Retrieved October 2, 2007)

Working Group Report 2003: Implementation of "Education & Training 2010" Work Programme. Working Group "Improving Education of Teachers and Trainers." Progress Report, November 2003. European Commission. Directorate-General for Education and Culture. Available at <http://ec.europa.eu/education/policies/2010/doc/working-group-report_en.pdf>

Working Group Report 2003: Implementation of "Education & Training 2010" Work Programme. Working Group 2ICT in Education and Training." Progress Report, November 2003. European Commission. Directorate-General for Education and Culture. Available at <http://ec.europa.eu/education/policies/2010/doc/it-technologies_en.pdf>

Recommendation of the European Parliament and the Council of 18 December 2006 on key competences for lifelong learning (2006/962/EC). Official Journal of the European Union L 394/10. Available at <http://eur-lex.europa.eu/LexUriServ/site/en/oj/2006/l_394/l_39420061230en00100018.pdf>

Proposal for a recommendation of the European Parliament and of the Council on key competences for lifelong learning (presented by the Commission). Brussels, 10.11.2005 COM (2005) 548 final. 2005/ 0221(COD) Commission of the European Communities. Available at <http://ec.europa.eu/education/policies/2010/doc/keyrec_en.pdf>

Key Competences for Lifelong Learning. A European Reference Framework, November 2004. Implementation of "Education and Training 2010" work programme. Working Group B. "Key Competences." European Commission. Directorate-General for Education and Culture. Available at <http://ec.europa.eu/education/policies/2010/doc/basicframe.pdf>

References

Bawden, D. (2001). Information and digital literacies: A review of concepts. *Journal of Documentation, 57(2)*, 218–259.

Buckingham, D. (2003). *Media education: Literacy, learning and contemporary culture.* Cambridge: Polity.

Castells, M., & Himanen, P. (2002). *The Information Society and the Welfare State. The Finnish Model.* Cambridge: Oxford University Press.

Díaz, M. G. (2006). The genesis and evolution of the EU eLearning initiative. In P. Ruohotie & R. Maclean (Eds.), *Communication and learning in the multicultural world.* University of Tampere and OKKA foundation.

Dighe, A., & Reddi, U. V. (2006). Women's Literacy and Information and Communication Technologies: Lessons that Experience Has Taught Us. Retrieved August 9, 2007 from http://www.cemca.org/CEMCA_Womens_Literacy.pdf

Fairclough, N. (2005). Critical Discourse Analysis. *Marges Linquistiques* 9, 76–94. Retrieved August 8, 2007, from http://www.ling.lancs.ac.uk/staff/norman/norman.htm

Foucault, M. (2000). Governmentality. In his *Power: Essential Works of Foucault 3: 1954–1984.* New York: New Press.

Freire, P. (1972). *Pedagogy of the oppressed.* New York: Continuum.

Gee, J. P. (2003). *What video games have to teach as about learning and literacy.* New York: Palgrave Macmillan.

Johnston, B., & Webber, S. (2006). As we may think: Information literacy as a discipline for the information age. *Research Strategies, 20*, 108–121.

Kellner, D., & Share, J. (2005). Toward critical media literacy: Core concepts, debates, organizations, and policy. *Discourse: Studies in the Cultural Politics of Education 3*, 369–386.

Lankshear, C. (1999). Literacy Studies in Education: Disciplined Developments in a Post-Disciplinary Age. In M. Peters (Ed.), *After the disciplines*. Westport, CT: Greenwood Publishing Co. Retrieved May 29, 2007, from http://www.geocities.com/c.lankshear/literacystudies.html

Lankshear, C., & Knobel, M. (2004). *A handbook for teacher research: From design to implementation*. New York & Maidenhead: Open University Press.

Lankshear, C. & Snyder, I. (2000). *Teachers and technoliteracy*. Sydney: Allen & Unwin.

Lemke, T. (2000). Foucault, Governmentality and Critique. Paper presented at the Rethinking Marxism Conference, University of Amherst (MA), September 21–24.

Levine, K. (1986). *The social context of literacy*. London & New York: Routledge.

Lewis, C. (2007). New Literacies. In M. Knobel & C. Lankshear (Eds.). *A new literacies sampler* (pp. 229–237). New York: Peter Lang.

Livingstone, S. (2003). The Changing Nature and Uses of Media Literacy. Retrieved January 24, 2007 from http://www.lse.ac.uk/collections/media@lse/pdf/media@lseEWP4_july03.pdf

Livingstone, S., van Couvering, E., & Thumim, N. (2008). Converging traditions of research on media and information literacies: Disciplinary, critical and methodological issues. In J. Coiro, M. Knobel, C. Lankshear, & D. Leu (Eds.), *Handbook of research on new literacies* (pp. 103–132). Mahwah, NJ: Lawrence Erlbaum Associates/Routledge.

Olssen, M. (2006). Understanding the mechanisms of neoliberal control: lifelong learning, flexibility and knowledge capitalism. *International Journal of Lifelong Education, 25(3)*, 213–230.

Popkewitz, T., Olsson, U., & Petersson, K. (2006). The learning society, the unfinished cosmopolitan, and governing education, public health and crime prevention at the beginning of the twenty-first century. *Educational Philosophy and Theory, 38(4)*, 431–449.

Porter, D., & Craig, D. (2004). The third way and the third world: Poverty reduction and social inclusion in the rise of 'inclusive' liberalism. *Review of International Political Economy, 11(2)*, 387–423. Retrieved May 5, 2007 from http://www.cevipof.mshparis.fr/rencontres/colloq/palier/Full%20paper/Craig,%20Porter%20FP.doc

Rodrigues, M. J. (2006). The European way to a knowledge-intensive economy—The Lisbon strategy. In Manuel Castells & Gustavo Cardoso (Eds.) *The network society: From knowledge to policy* (pp. 405–424). Washington, DC: Johns Hopkins Center for Transatlantic Relations.

Roelvink, G., & Craig, D. (2004). The man in the partnering state: Regendering the social through partnership. Research Paper 13. Retrieved October 2, 2007 from http://www.arts.auckland.ac.nz/lpg/Researchpaper13.pdf.

Scribner, S., & Cole, M. (1981). *The psychology of literacy*. Cambridge, MA: Harvard University Press.

Simons, M. (2006). Learning as investment: Notes on governmentality and biopolitics. *Educational Philosophy and Theory, 38(4)*, 523–540.

Simons, M., & Masschelein, J. (2006). The learning society and governmentality: An introduction. *Educational Philosophy and Theory, 38(4)*, 417–430.

Street, B. (1984). *Literacy in theory and practice*. Cambridge: Cambridge University Press.

————. (1995). *Social literacies: Critical approaches to literacy development, ethnography and education*. Essex: Longman.

————. (2003). What's "new" in new literacy studies? Critical approaches to literacy in theory and practice. *Current issues in Comparative Education 5(2)*, 77–91.

————. (2004) Understanding and Defining Literacy: Scoping Paper for EFA Global Monitoring Report 2006. UNESCO: Paris.

Suoranta, J., & Vadén, T. (2007). From social to socialist media: The critical potential of the wikiworld. In P. McLaren & J. Kincheloe (Eds.), *Critical pedagogy: Where are we now?* (pp. 143–162). New York: Peter Lang.

Tornero, J. M. P. (2004). Promoting Digital Literacy. Final report EAC/ 76/ 03. Retrieved November 25, 2007, from http://ec.europa.eu/education/programmes/elearning/doc/studies/dig_lit_en.pdf

Tuschling, A., & Engemann, C. (2006). From education to lifelong learning: The emerging regime of learning in the European Union. *Educational Philosophy and Theory 38(4)*, 451–469.

Uzunboylu, H. (2006). A review of two mainline e-learning projects in the European Union. *Educational Technology Research & Development, 54(2)*, 201–209.

Williams, R. (2005/1974). *Television*. London & New York: Routledge.

Youngman, F. (2000). *The political economy of adult education & development*. London & New York: Zed Books.

CHAPTER SIX

Digital Competence— From Education Policy to Pedagogy:

The Norwegian Context

MORTEN SØBY

> The prosperity of a nation, geographical region,
> business or individual depends on their ability
> to navigate the knowledge space.
> (Pierre Lévy)

Digital Knowledge Promotion Reform?

The implementation of the Knowledge Promotion Reform has meant that digital competence plays an important role in the Norwegian education system. Perhaps the best example is the use of digital tools, which is defined as a basic skill in the curriculum. This makes Norway the first country in Europe with a curriculum based on digital skills. In white paper no. 30 (2003–2004) *Kultur for læring (A learning culture)* digital competence is defined as:

> [. . .]the sum of individual ICT skills, such as reading, writing and maths, and more advanced skills ensuring a creative and critical use of digital tools and media. ICT-

skills include making use of software, searching, finding, processing and controlling information from various digital sources, while critical and creative ability also requires ability to evaluate information and sources, interpretation and analysis of digital genres and media types. Thus, digital competence can be regarded as a very composite form of competence. (2003–2004, p. 48)

Digital competence is a multimodal and complex concept constantly changing with the development of digital media. Media development is multidisciplinary by its very nature. In the space of only a few years, digital competence has established itself as a key concept in education policy and in educational research. This chapter will touch upon the history of the concept that has emerged from the tension between education policy and educational research. What is digital competence? Different definitions reflect different positions in the current debate. What challenges arise for education policy and educational research?

National Plans for ICT in Education

In the Norwegian Knowledge Promotion Reform, digital skills have been ascribed the status of being the fifth basic skill. Being able to use digital tools is defined alongside other basic skills, such as reading, writing, basic mathematics and using the spoken word. Ability to use digital tools has been included in the competence targets for all subjects on all levels—albeit to different degrees depending on which curriculum one is following. The digital Knowledge Promotion Reform in the curriculum is also based on national action plans: In the action plan for *IT in Norwegian Education 1996–99* the implementation of technology occupies a central position. The next action plan for IT in education (2000–2003) pursues the challenges associated with the implementation of IT in education. At the same time, this plan prioritizes the development of schools and maintaining a comprehensive perspective regarding the academic and pedagogical use of IT.

A series of projects and activities carried out in 2000–2003 helped create a basis for the subsequent *Program for digital competence 2004–2008*. The program further pursues a comprehensive framework for school development and implementation of ICT from the previous action plans. However, the program also introduces new, ambitious national targets and priorities set through the vision "digital competence for everyone." The program deals with how information and communication technology (ICT) influences the quality of education, incentive for learning, forms of learning and learning outcomes. The program has four main objectives:

- Norwegian educational institutions should have access to high-quality infrastructure and services in 2008
- digital competence will be central to education at all levels in 2008
- the Norwegian education system should be among the best in the world in this area by 2008
- ICT should be an integral policy instrument for innovation and quality development in Norwegian education in 2008.

The Ministry of Government Administration and Reform has had the responsibility for developing the eNorway strategies. The Ministry of Government Administration and Reform is responsible for the Government's administration and personnel policy, competition policy, national policy for development and coordination of the use of information technology and measures to make government more efficient and service-oriented. Digital competence is a central concept to achieving innovation and modernization in *eNorway2009* (MOD, 2005). The concept of digital competence is taken further in white paper no. 17 (2006–2007) *An information society for all*. It includes the following statement:

> The government espouses the objective of achieving a modern education system with an active and discerning approach to new technology and which draws on the potential that exists in the interface between digital youth culture and the schools' more traditional learning culture. (2006–2007, pp. 28–29)

Digital competence has set the agenda for innovation, education and pedagogy. The concept has had a double function as an agenda setter. On the one hand, it is the principal political concept in innovation policy and in the educational reform: The Knowledge Promotion Reform. On the other hand, the concept has become an objective in the development of the schools and in practical pedagogy. Educationists are now working on anchoring digital competence in theories for learning and media development and further developing the concept. The term digital competence has been something akin to a password into new fields politically as well as pedagogically.

Password into New Fields?

From the classical period until today, intellectual trends have had their institutions: the courts, the salons, the newspapers and the journals. It is in such institutions that new concepts and terms are set in circulation, with great speed and motion. New terms sum up the present time, crystallize trends and create new ideas and visions. Such terms make it possible, for a time, for users to create for themselves a separate discussion area. Some terms can function as the "word in vogue" of the moment and live a short life as a buzzword. Other terms can create a lasting trend and function as passwords to a new field. Passwords generate ideas, contribute new ways of thinking and provide access to discussions. Digital competence may be the password into a new multi-disciplinary research area, the guide in a process of lifelong learning and to objectives in education policy.

Digital competence has established itself as a collective term for understanding the complex connection between individuals, organizations, ICT and society. The concept is increasingly central to research, education policy, learning and societal debate. In the report *Digital skole hver dag (digital schooling)* digital competence is defined as" . . . skills, knowledge and attitudes required by everyone in order to be able to use digital media for learning and mastery in the knowledge society" (ITU, 2006, p. 8).

Digital competence can be seen as a concept whose status is "essentially contested" (Connolly, 1993). It has a vague conceptual core or essence that is subject to discussion on a fundamental level. Much in the same way as with the word "democracy," several participants will join discussions and efforts to define the concept of digital competence. A discussion of digital competence may take place along three dimensions. First, it is about appraisal or values. Second, there is a complex span between skills and knowledge and formative education. Third, there is an openness that creates potential for several possible interpretations and areas of use.

The discussions of the terms *digital skills*, *digital competence* and *digital "bildung"* are numerous and complex in Scandinavian public debate. Use of the various terms in policy documents on education policy shows that there is an ongoing debate and different interpretations within both educational science and politics. In the new school curricula in Norway the term digital skills is connected to the use of digital tools. In the main report of The Committee for Quality in Primary and Secondary Education in Norway (NOU 2003, p. 16) *In the first place (I første rekke)* and in the *Program for digital competence* (Ministry of Education and Research, 2004) there is a broader understanding of digital competence in the sense of digital formative education. Furthermore, the ongoing debate about these terms is linked to a dynamic and rapid development of convergent digital media.

This chapter is based on the thesis that there is unexploited potential for learning outcomes associated with the academic/professional and pedagogical use of digital media. In other words, its premise is that digital media are not used to their full potential in institutions for learning as of today. In the years ahead, the development of digital media will create new opportunities, but also barriers to implementation and innovation with regard to learning outcomes. Internet and mobile services make communication richer, spanning more media and more personal uses. The digital arena is just as much a place for differentiation and cultural diversity as a driving force for homogeneity. This provides previously unknown and novel opportunities for learning. However, for it to be successful, the pupils have to be included. Meanwhile theories for learning have to be updated and come into step with the digital revolution. Pedagogical practice is still dominated by book technology. Some educational researchers even regard ICT as a threat.

Cultural Technologies

In a historical perspective, technology is often perceived as a threat before it is incorporated into culture. In cultures based on the spoken word, writing has often been regarded with skepticism and characterized as unnatural and inhuman. Plato *(The Phaedrus* dialogue) feared that writing would be produced outside of consciousness and destroy the memory. Meanwhile, the art of writing has become completely natural to us. Gutenberg's controversial printing press has been implemented in today's schools. Book print is natural within the schools and is no longer viewed as technology.

Ong (1982) shows in *Orality and Literacy* that writing and books are also technology: "Technologies are not mere exterior aids but also interior transformations of consciousness [. . .] Writing heightens consciousness. Alienation from a natural milieu can be good for us and indeed is in many ways essential for full human life" (Ong, 1982, p. 82). According to Ong, writing becomes interiorized. That makes it difficult to see writing as technology. There is a close connection between the philosophy of the Enlightenment and printing techniques. For example, in seeing a book's print as "natural"—something that has lost its technical character—pedagogy has forgotten how technology and culture are interwoven (Søby, 1998).

In the pedagogical classic *Emile*, Rousseau (1962) warns against providing children with access to globes and maps. Rousseau considers children to be incapable of navigating by means of using maps. He is skeptical about this technology and argues that education and upbringing should take place in natural surroundings. Maps are good examples of compressed representations of our surroundings that have been developed over many thousands of years. Maps are cognitive prostheses, which we can learn to use in order to navigate. Today's teacher training in Norway is still characterized by the attitude that it is pedagogically and politically correct for children to learn to write using a pencil rather than using a keyboard and word processing. The basis for established pedagogical theory and practice is anchored in oral and written culture. At the same time, the internet is a natural part of children's and young people's upbringing.

Snow warned in the 1950s against separating technology and culture. This would lead to technology developing into a form of rationality with a basis in science without cultural knowledge and that cultural analyses in the fields of the humanities and social science lacked technological knowledge (Snow, 1959). With the concept of digital competence, the challenge is to have a combined discussion of technology, culture and pedagogy. Manuel Castells contributes to any such discussion. He describes the development of knowledge through:

> [...] a co-evolution between the human brain and the computer learning from each other [...] So a computer cannot become a subject in its own right, but I could have a computer as an extension of the mind, whose reactions and help affect the mind, inducing individualised co-evolution between people and their machines. (2003, p. 137)

Castells describes ICT as a cognitive prosthesis, a perspective shared by Marshall McLuhan. In the 1960s, McLuhan described the media as an extension and perfection of the human senses. In his view, the electronic media are prostheses—a global extension of the body and the brain doing away with both time and space:

> Rapidly, we approach the final phase of the extensions of man—the technological simulation of consciousness, when the creative process of knowing will be collectively and corporately extended to the whole of human society, much as we have already extended our senses and our nerves by the various media ... Any extension, whether of skin, hand, or foot, affects the whole psychic and social complex ... (McLuhan, 1968, p. 19)

According to McLuhan, mankind overcomes its natural limitations by using the media as prostheses. Gregory Bateson's example about the blind man and the cane may illustrate this point. When the man has learned how to use it, it becomes a part of the hand. The hand is part of the body which interacts in a complex system. For Bateson "a mind"—the mental or the psychological—is an aggregate of parts that interact (Bateson, 1979, p. 102). The term metaphor comes from the Greek meta-pherein, which means "transfer" or, more correctly, "carry to another place." The meaning of the metaphor does not lie in one system of references or another, but in the interaction between them. The metaphor can give insight because it is ". . . . our means of effecting instantaneous fusion of two separate realms of experience into the one illuminating, iconic, encapsulating image" (Nisbet, 1969, p. 4).

The use of the metaphor has long been associated with poetic creativity, subjective characters and ornamental rhetorics. Nevertheless, there is a tendency for the metaphor to be viewed as far from decorative and not estranged from thought and action. Research is underway regarding the extension and the importance of metaphors in everyday life. Another approach studies the role of the metaphor in the world of science. At the same time, through the work of historians, sociologists and anthropologists we have moved away from the supposition that researchers are hermetically closed in their laboratories, extracted from social and cultural activities. The metaphor is already there: Metaphor circles around in town; it freights us as its inhabitants, follows all kinds of routes, with street corners, red lights, one-way streets, crossroads or crossings, speed limits and pleas. Within this communication means we are— metaphorically speaking, of course, and in relation to a way of living—contents and wording: passengers, preoccupied and transferred by the metaphor.

When children are playing a computer game, the game is part of the mind. Or, in the words of Castells, the "internet is the fabric of our lives" (Castells, 2001, p. 1). A pupil at school today has to master different meeting venues; from the intimate chat sphere, where pupils have daily chats about boyfriends/ girlfriends or Norwegian essays, to other, much larger arenas for role-playing games over the internet for months along with several thousand people. It is important to see one's own role in different contexts. But it is also important to use the internet to expand your learning horizon and ultimately to be creative and invent yourself on the internet through blogging or podcasting. Seeing technology as a cognitive prosthesis contrasts with political and pedagogical perspectives characterized by metaphors about tools and an instrumental view of knowledge.

In some school development projects, acquisition and use of LMS[1] email are defined as the final goal for completing the digitalization of the school. This understanding does not take into account young people's true usage of the media and their forms of communication, nor does it contribute to modernization of pedagogy in practice. Over the last three years, both the internet and popular internet-based forms of communication have changed, or gained widespread popularity among children and young people. A new generation of web-based services makes it possible to cooperate and share information online. The user experience is, in many cases closer to that of the local PC than ordinary websites. Web 2.0 opens the way for mass publication (web-based social software), such as via blogs and wikis. Blogging is particularly interesting with regard to the school, as a text-based, low technological solution with the emphasis on statements of opinions and comments.

Norwegian education policy is characterized by two extremes: traditional pedagogy and progressive pedagogy. Traditional pedagogy criticizes the schools of today for poor knowledge and poor knowledge transfer. They say the teacher's role and authority must be strengthened and more discipline is required in the classroom. Subject targets in the curriculum must be unambiguous and examinations must show if targets have been achieved. Progressive pedagogy emphasizes the pupils' independent activities, participation in the learning process and project work. This polarized view is, of course, a simplified one, but it provides contrasts that help place the pedagogical understanding of technology in context. In the classroom, many teachers are reticent about making use of their pupils' digital skills academically and pedagogically. The skepticism towards technology is evident not only in pedagogy in practice, but also in pedagogical theories.

The Hidden Syllabus of Pedagogy

Both traditional and progressive pedagogy are based on Age of Enlightenment ideals and printing press technology. These historical roots have reduced the development of technological knowledge in pedagogy. Traditional pedagogy has taken an instrumental approach to technology, viewing it as a tool for effective learning and teaching. One example is the emergence of *Computer Assisted Instruction* (CAI) and *Computer Based Training* (CBT) in the 1980s. The progressive camp of pedagogy at the time criticized the instrumentalist approach and the behaviorist theory of learning behind CAI and CBT. Such software for repetitive practice has been developed for skill testing, and it had and still has a low degree of interaction. The criticism from the progressive end of pedagogy of both educational reforms and CAI/CBT was relevant. However, it failed to provide an alternative view of technology to any meaningful degree. The perception remained that a combination of technology and pedagogy always resulted in prefabricated learning packages such as CBT/CAI.

The criticism of the instrumentalist approach presented from the standpoint of progressive pedagogy was inspired by Paulo Freire and Jürgen Habermas. The criticism from educational researchers in Scandinavia charged that the educational technology of the 1970s would lead to a mechanical materialization of information dissemination and qualification in a behavioral segment through a technocratic production process: Educational technology improves adaptation effectively.

The criticism of educational technology in the 1970s was mainly directed against making the education system more technocratic, with traditional didactics representing a means-to-an-end rationale. Critical pedagogy emerged under terms like dialogue pedagogy, project orientation, problem orientation and participant-governed learning. A common denominator for the critical branches of pedagogy was the emphasis on communication and cooperation: that is, an interactive, alternative pedagogical system.

However, implicit in the criticism that education policy is controlled by technocrats is the sentiment that technology is harmful to mankind, that technology can't expand our cognitive capacity at all and that we cannot use it for reflection or to expand our horizons. The progressive pedagogical system is based on the ideals of the Age of Enlightenment and printing technology. When this is combined with a humanistic orientation, pedagogy becomes a perceived defense against technology. Progressive pedagogy has thus only to a very small extent developed technological knowledge and the terms needed to understand digital media.

In Norway, this contributed to making technophobia a hidden part of the syllabus in pedagogical teaching. In practice, this sentiment is still alive in the form of skepticism towards technology amongst some teachers. Many teachers are reluctant to use their pupils' digital skills in the classroom, academically and pedagogically. This skepticism towards technology in practical pedagogy can in many instances be traced back to progressivist pedagogical theories.

Many researchers of progressive pedagogy are still closet technophobes. This is paradoxical since information technology and learning converge in the current multi-disciplinary trend: *computer-supported collaborative learning*

(CSCL) and situated cognition and a learning community. The objective of the alternative pedagogical theories is exactly what is emphasized in computer-supported collaboration: digitalization advances dialogue, collaboration and problem orientation. According to Timothy Koschmann (1996, p. 3) there have been four paradigm shifts during the development of the technology of teaching and learning:

- *Computer-Assisted Instruction* (CAI): Starting from a behaviorist theory of learning and repetition.
- *Information Processing Theory* (ITS): Attempts to develop artificial intelligence systems for transfer of information based on an instrumentally anchored view of knowledge.
- *Logo-as-Latin:* Cognitive constructivist theory of learning. Papert's development of Piaget has given us Logo-programming and *Lego Mindstorms.*
- *Computer-supported collaborative learning*: Socio-culturally oriented theory of learning

Koschmann believes these four directions of thought about educational technology constitute paradigm shifts in Kuhn's sense. In the context of discussing paradigm shifts in the natural sciences, he argues that "the shifts that have occurred in IT were in fact driven by shifts in underlying psychological theories of learning and instruction" (Koschmann, 1996, p. 3). In this argument he seems to assign exaggerated power to theories of teaching, and within his arguments about how computer-supported collaborative learning is anchored and constructed socially Koschmann's views of information technology appear unduly instrumental. He uses the metaphor of tools throughout, in the manner of tools being used that are somehow external to learning processes.

It is tempting to turn Koschmann's argument on its head. Is it the development of information technology that is fundamental? In that case, the stages of educational technology have been generated by the development of information technology rather than the development of theories of learning. Are CAI and ITS pedagogical products of the contemporary mainframe computers, terminals and computer experts in white coats? Has *Logo-as-Latin* emerged due to advances in programming languages and the spread of the PC? Did the internet lead to renewed interest in project-oriented, problem-oriented, collaboration-oriented, situated pedagogy and so forth? Do these different trends re-emerge and become united in CSCL? Perhaps the theories of learning and educational technology are products of the current information technology at any given time?

Østerud (2004) argues that ICT may be the midwife for a new pedagogical system, a third way or a synthesis between progressive *Bildung*-oriented pedagogy and a restorative knowledge-oriented pedagogy: ICT makes way for a new school model for the 21st century and the learning of the future. That doesn't mean that ICT will on its own automatically create innovation and new spaces for learning. The potential of digital media can only be realized if it is anchored in a pedagogical, social and organizational context, supported by political commitment. That is why it takes time to realize the learning benefits of ICT in school development projects. Utilization of ICT in central learning activities depends on the school facilitating the use of ICT in a comprehensive way (ITU, 2006). This means clearly defined pedagogical targets, professional ICT infrastructure, school leadership, organizational development and competence building.

Digital Competence—Bricks in a Development of Concepts

A review of the development of the concepts involved in digital competence shows that they have several origins. The concepts range from "computer operating licence" skills to digital competence and a digital *Bildung* and are frequently used with different meanings in different policy documents on education.

There isn't a clearly defined international frame of reference for this field. Three different trends can be highlighted: one is associated with the definition of basic skills within ICT, such as word processing, spreadsheets, presentations software and internet searches. Another is associated with concepts such as the fourth basic skill and the fourth cultural technique, which are about fundamental ICT skills as a basis for professional use. A third is based on an updated concept of educational *Bildung* with the focus on broader digital competence and expertise.

Digital competence is related to *ICT literacy*[2] and *digital literacy*. These two terms appear in different contexts and under various definitions. They exist in a new multi-disciplinary research field that to a certain extent is based on *media literacy, media studies, and media education*. The concepts are also used in popular science discussions and in mass media (cf. Gilster, 1997). They also appear in policy documents on education published by the OECD, EU and as part of national action plans for ICT in education (cf. New Zealand and Singapore).

Traditionally, *literacy* in English literature has been regarded as basic skills in reading and writing independent of social context. Recent *literacy* research has extended the meaning of *literacy* to include the writing technology in social and cultural practice. In Kathleen Tyner's classic *Literacy in a Digital World*, two definitions of *literacies* are introduced (Tyner, 1998). Tyner distinguishes between *tool literacies*, which are concrete and relate to using computers, networks and media technology, and *literacy of representation*, which is about understanding how media types are organized, what they represent and how they create meaning.

Internationally, numerous definitions of *literacy* exist. Even if the term and concept of *literacy* originates in the culture's literary field, it exists with a series of prefixes: *media-, technology-, visual-, computer-, information-* and *multimodal*, etc. David Buckingham uses the term in the plural as *multiple literacies*:

> The increasing convergence of contemporary media means that we need to be addressing the skills and competencies—the multiple literacies—that are required by the whole range of contemporary forms of communication. Rather than simply adding media or digital literacy to the curriculum menu, or hiving off information and communication technology into a separate school subject, we need a much broader reconceptualisation of what we mean by literacy in a world that is increasingly dominated by electronic media. (Buckingham, 2006, p. 275; see Chapter 4 here)

The first broad presentation of digital competence in the Norwegian public space is made in ITU's report *Digital kompetanse: fra 4. basisferdighet til digital dannelse (Digital competence: from 4th basic skill to digital bildung)* (Søby, 2003). ITU's report on digital competence (2003) is inspired by work by the *Educational Testing Service* (ETS)[3] in the U.S.. ETS put together an international panel in 2001 to study the relationship between literacy development and ICT. The panel was made up of experts, policymakers and researchers from Australia, Brazil, Canada, France and the U.S.. In the report *Digital Transformation. A Framework for ICT Literacy* the term *ICT literacy* is defined as follows:

> ICT literacy is using digital technology, communications tools, and/or networks to access, manage, integrate, evaluate, and create information in order to function in a knowledge society. The panel's definition reflects the notion of ICT literacy as a continuum, which allows the measurement of various aspects of literacy, from daily life skills to the transformative benefits of ICT proficiency (Educational Testing Service, 2002, p. 2).

ETS identifies five critical components that represent a set of skills and knowledge. The report emphasizes that this set is part of the development of increasing cognitive complexity: from simple skills to meta-cognition and expert knowledge:

- **Access:** Knowing about and knowing how to collect and/or retrieve information.
- **Manage:** Applying an existing organizational and classification scheme.
- **Integrate:** Interpreting and representing information. It involves summarising, comparing and contrasting.
- **Evaluate**: Making judgements about quality, relevance and usefulness, or efficiency of the information.
- **Create:** Generating information by adapting, applying, designing, inventing or authoring information (Educational Testing Service, 2002, p. 3).

The report argues that *ICT literacy* should not be defined primarily as mastering static and technical skills. One important prerequisite is basic competence (reading, writing and arithmetic) as well as the ability for critical thought and problem solving. The ETS report also notes that *ICT literacy* will be a continually changing concept.

Starting from the concept of *ICT literacy* the report suggests a review of school curricula to adapt them to different levels of skill and age. Equally, new ways of assessment and digital folders are recommended to measure and document the level of digital competence. According to the ETS report, innovation within the education system based on *ICT literacy* will be an essential factor in economic growth, and digital competence is necessary in order to function in the information society.

Digital literacy involves the ability to develop the potential inherent in ICT and use it innovatively for learning and for work. This requires a certain level of confidence with digital media and is considered a key concept for lifelong learning. The concept of *digital literacy* has a central place in several of the EU's research and education programs. In the *eLearning program* from 2003, *digital literacy* is justified on the basis of being part of e-citizenship:

The ability to use ICT and the Internet becomes a new form of literacy—"digital literacy." Digital literacy is fast becoming a prerequisite for creativity, innovation and entrepreneurship and without it citizens can neither participate fully in society nor acquire the skills and knowledge necessary to live in the 21st century. (European Commission for Education and Culture, 2003, p. 3)

In attempting to set the agenda for education in the 21st century, the European Parliament and the Council of the European Union has stated that digital competence is one of eight key competences for lifelong learning. Through their recommendations, learning is not only understood as a lifelong endeavour but it is also recognized that the formal education systems provide only a subset of all different settings where learning and development occur. According to the EU policy this new competence is important both at school and outside school. EU policy documents list the following skills and competencies: downloading, searching, navigating, classifying, integrating, evaluating, communicating, collaborating and creating. In terms of education policy, digital competence has become an essential concept in Europe.

Skills and Basic Competence

Several international studies note that talking in terms of skills provides only a narrow perspective on education and learning activities. The OECD invited its member countries to participate in a four-year project: *The Definition and Selection of Competencies* (or, DeSeCo; OECD, 2002), which originated in an increasing international interest in outcomes and the effect of training and education, as well as a need for a common frame of reference for identifying and analysing so-called basic components. Competence is defined here as: "[. ..] the ability to meet demands or carry out a task successfully, and consists of both cognitive and non-cognitive dimensions" (OECD, 2002, no page).

DeSeCo focuses on three basic competency categories (see Figure 6.1). These competencies are important in different life situations and are defined as necessary to all of them. The DeSeCo report emphasizes that basic components must be selected and defined in accordance with what societies and individuals within particular societal groups and institutions value.

The DeSeCo report has become the foundation for international collaboration on work related to the concept of competence. The use of the concept of competence in connection with basic education is relatively new. The concept of competence has been applied to adults' knowledge and skills. With regard to lifelong learning, a comprehensive concept of competence has become an important term in education policy, planning and quality studies.

Figure 6.1: The Categories of Basic Competencies (from OECD, 2002, no page).

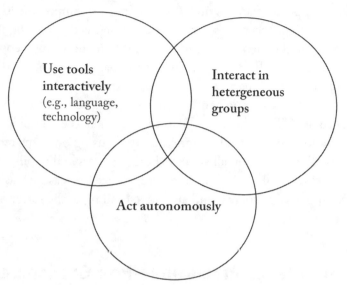

OECD statements emphasize that building competence concerns the whole person. It is about relating proactively to challenges posed by the environment and times in which we live, along with meeting highly complex demands. Mere knowledge and skills are not sufficient in themselves. Strategies, attitudes and procedures are also required. Competence is a performance-related term describing a preparedness to take action:

> Competence is the ability and readiness to meet a challenge through action, when it is often implicit that the challenge is not a given, but depends on context; that it is not a routine challenge, but novel and not judged by given criteria for success, but by the outcome which form is not known in advance. (Hermann, 2005, p. 9)

The OECD's view of competence influences the main report of The Committee for Quality in Primary and Secondary Education in Norway, *I første rekke (In the first row)* (NOU, 2003, p. 16). Kvalitetsutvalget (The Committee for Quality) emphasizes that basic education must focus on increasing the pupil's basic competence beyond its current level. The committee has defined social competence and the development of learning strategies as basic skills. The Committee proposed that digital competence must be given a concrete form and be built into curricula for the subjects and ICT is described as part of a collective development strategy. Digital competence is regarded as having equal weight with reading, writing and arithmetic skills and is part of an integrated perspective encompassing learning strategies and social competence.

Digital competence and digital *Bildung* are more important in the information society than focusing solely on skill-based activities. Digital *Bildung* expresses an overall understanding of how children and young people learn to develop their identity. The term will also encompass and combine the application of skills, qualifications and knowledge. In this way, digital *Bildung* points to an integrated and comprehensive approach that enables us to reflect on the influence of ICT on different qualifications such as communication skills, social skills and pupils' critical judgments. By focusing on a greater degree of the use of ICT integrated in all subjects both teachers and pupils will develop the necessary ICT skills while building competence in areas such as navigation and critical appraisal of sources and an understanding of the social significance of digital technology.

Knowledge Promotion—New Curriculum in Norway

Public consultation for the new school curricula in the Knowledge Promotion Reform (*Kunnskapsløftet*) discussed the terms digital skills and digital competence. Meanwhile, there has been conflict over the extent to which digital competence should dominate the curriculum.

There was also conflict over fundamental digital skills being defined at the same level as other fundamental skills, such as reading, writing, arithmetic and oral communication. The aims of the curriculum require the use of digital tools in individual subjects. In the social sciences, fundamental digital skills are described as follows:

> Being able to use digital tools in social science subjects includes making estimates, searching for information, exploring websites, critical appraisal of sources, having good Internet sense and selecting relevant information on academic topics. Digital skills also involve being aware of the protection of privacy and intellectual property rights and applying and adhering to rules and norms for Internet-based communication. Using digital communication and collaboration tools involves preparing, presenting and publishing multimedia products individually and in common with others, communicating and collaborating with pupils from other schools and countries. (Norwegian Directorate for Education and Training, 2006a, p. 120)

The presentation of digital tools in social studies is part of the concept of digital competence. This shows that the Knowledge Promotion Reform does not use the terms precisely and that the text refers to simple digital skills and broad digital competence interchangeably.

Some more examples: According to the science competence targets for year 7, pupils are supposed to know how to: "publish results of their own investigations by using digital tools" (ibid., p. 87). The digital tools in question include surveying (opinion polls) tools, to which many have access through LMS, searching and information gathering on the internet, email and using digital technology such as data loggers. The various publication options include internet journals, wikis, blogs, and the websites nysgjerrigpermetoden.no and miljolare.no. Another competence target for year 7 is: "making relevant weather measurements and presenting the results using digital tools" (ibid.). In practical terms, these may be weather data measured with digital tools such as data loggers or digital thermometers. Common presentation programs are PowerPoint in Microsoft Office, Impress in Open Office and Keynote from Apple.

As part of their Norwegian studies in the second year, pupils have to use a "computer to create text" (ibid., p. 44), while in 4th year they have to "perform information searches, creating, storing and retrieving texts using digital tools" (ibid., p. 45). After the 7th year the pupils have to "use digital writing tools in an authoring process and for the production of interactive texts" (ibid., p. 46). In secondary school the pupils have to work with multimodal texts via digital media for their project study for the general studies qualification.

Growing use of digital tools both at school and outside school provides great opportunities for children and young people to produce rich, multimodal texts. The traditional printed and spoken texts encountered at school can and should be supplemented by the pupils' abilities to create their own multimodal texts. Digital competence involves interpretation, and the reading and writing of digital media. In other words, this involves pupils in the production of their own multimodal texts. This dual approach is actually a key to understanding the trend towards web services that are more user-driven and interactive. This new type of web services depends on active participation from the users. This includes everything from blogs to wikis, podcasts/vidcasts and social networking services. Services like these are popularly called *Web 2.0*.

Liestøl (2006) argues that schools must improve their ability to make the unique competence of pupils and teachers more visible. Teachers possess valuable knowledge based on traditional media, while pupils have experience with and competence in new digital media. Often the thematic content overlaps. Literary and historical texts can often be found in digital form, for example, as films and computer games. In project work, teachers and pupils can develop digital competence together if both parties supply their unique skills and knowledge that can be developed further with digital media. ITU Monitor 2005 (Norwegian survey) shows that teachers who facilitate a great deal of collaborative work are more inclined to employ varied methods of teaching and assessment. They also collaborate more often with colleagues at the school and outside and are more inclined to employ new technology for their teaching. The teachers are proactive in how they see their own role and the use of working practices and technology in order to facilitate better quality of work to achieve learning and establish identity. This requires a high degree of inclination towards critical awareness and digital competence.

An example from the school: *Blokka* is a writing project in the 5th year at Eberg primary school in Trondheim. The main objective is to instill a joy of writing in the children, and computers are used in the writing process. The work is characterized by collaboration, and the pupils learn how to use a computer program to create websites with the emphasis on graphic design, pictures and hyperlinks. The teacher shows the pupils an empty apartment block awaiting people to move in, graphically presented on the web. The block has many apartments, and the pupils are supposed to describe the residents. Then the teacher introduces a mysterious event. The pupils continue working on new texts based on the teachers' story and the accounts given by the other pupils. They are finished when everyone has written their own ending. The pupils have to take into account the main course of events in the narrative and read many texts while working on it. This is good practice for both writing and reading. The project involves advanced text management and a complicated writing process, actively using pictures, tables and other visual and rhetorical aids. This adds value to the writing process, with respect to writing and creativity but also with respect to integration of ICT in the process. The teacher manages to create motivation and instill enthusiasm in the pupils. The production of text and collaboration are essential (ITU, 2007a).

Digital media represent new opportunities for schools to access updated sources, but developing critical faculties requires time and knowledge for teachers and pupils alike. The school's aim is to encourage pupils to be critical of and question information they find on the internet. Being critical of sources is about appraising the quality of the information one gathers with regard to the questions one wants answered. Jenkins (2007) emphasizes that young people don't just respond to existing digital sources but contribute to new digital content. Jenkins argues that young people, therefore, have to be trained to develop a critical attitude to the ethical choices they make, both as participants and as communicators of digital content. This is particularly important because of the potential effect on other people of what they publish.

For schools, the use of digital sources is a pedagogical challenge and not a computer technology-related problem. The solution lies in focusing on the learning process and not the product. The Norwegian curriculum also focuses on critical appraisal of sources and critical thought when using digital media. The subject of KRL (Christian Knowledge and Religious and Ethical Education) stresses the importance of being able to use material available digitally—pictures, text, music and film—in ways that combine creativity with critical awareness and appraisal of sources.

Åskollen primary school provides an example. At Åskollen school, the 6th and 7th year pupils were working on a project with the aim of producing multimodal texts about the Drammen municipality. The work was organized in groups, and each group had to make their own multimodal text. There was an emphasis on how to find information in today's world, which was part of the pupils' groundwork when assessing various digital sources themselves. The pupils selected the sources themselves but got help from the teacher in deciding if what they found was useful, and in what way it was useful. The pupils worked in groups and all together finding background information about their chosen topics. The various topics led to many classroom discussions in which the pupils and the teacher together considered methods of assessing digital content. Intellectual property rights to digital sources also became a central theme of discussion. After approaching the Mayor of Drammen municipality, the pupils were given access to the image library of the municipality (ITU, 2007b).

Development of practical digital competence poses a series of challenges for school management, teachers and pupils: How should one work with digital tools in different subjects on different levels? What, for example, are the consequences for learning if pupils should be able to use particular digital software and tools to support interactive learning simulations and exploration? It means that measurement instruments, graphic calculators, PDA, mathematical modeling software and web-based resources have to be integrated into the learning process. In all, the Knowledge Promotion Reform is the start of an extensive development of schools, which requires co-ordinated follow-up in education policy.

It is important to include the general studies part of the curriculum in this development. It places the emphasis on the general knowledge perspective through which pupils should be stimulated to develop into creative people: "The aim of the education is to expand the ability of children, young people and adults for comprehension, experience, empathy, expression of self and participation" (The Norwegian Directorate for Education and Training, 2006, p. 3). The curriculum includes several statements of objectives that refer to providing a formative education that promotes good general and cultural knowledge. This can contribute to an updated and broad concept of digital competence and a vision of digital *Bildung*.

The State of Digital Competence in Basic School and Upper Secondary School

To meet the challenges of a complex and rapidly changing information society we have to develop a digital learning culture. To do this requires co-ordinated effort and ICT infrastructure. White paper no. 17 (2006–2007) *Eit informasjonssamfunn for alle (An information society for all)* states that Norway shall be a pioneering nation in the use of ICT in education, and it stresses the need for more investment in ICT in the education system. Important aspects of this are to ensure better access to PCs and the internet for all pupils and teachers and to increase the emphasis on digital teaching resources. Competence is society's most important resource and a dominant factor in value creation, economic growth and the development of society.

The report from The Norwegian Directorate for Education and Training entitled *The State of Equipment and Services in Education 2006–2007* shows that PC availability is better since the ratio of number of pupils per computer in compulsory school was reduced from 6.5 in 2005 to 4.7 in 2007 (The Norwegian Directorate for Education and Training, 2007, p. a:5). The corresponding figures for upper secondary school level are 2.5 and 1.8. Over 90 per cent of all computers in compulsory school are connected to the internet compared to 80 per cent two years ago. On average there are 4.2 lower secondary school pupils and 6.1 primary school pupils per computer with an internet connection available for use by pupils. For upper secondary education, 96 per cent of computers are reported to be connected to the internet, i.e., each school has on average about 1.9 pupils per networked computer (ibid., pp. 5–6). Even if this is an improvement with respect to access to ICT, the figures show that the infrastructure, internet access and bandwidth are not sufficient to fulfill the ambitions for use of digital tools in the curriculum in the Knowledge Promotion Reform.

ITU Monitor is a longitudinal study to survey digital competence in basic education. ITU Monitor is the only survey that provides a representative picture of how and to what extent ICT is used pedagogically and in specific subject areas in Norway. In addition, the study provides insights into the organizational aspects of ICT use in schools, such as planning, leadership, technological infrastructure and professional development among teachers.

ITU Monitor 2005 reports that teachers and pupils use home computers for a lot of school-related work that is not reflected in the pedagogical practice of the schools. It reports that many pupils develop more varied forms of digital competence at home than they are able to do at school (ibid., p. 82). At school they primarily use the internet and text-based services, while activities related to communication, games, multimedia, downloading of software and use of other equipment such as digital cameras and mobile services are not used at school beyond a negligible degree. The difference between what they use at school and what they use at home is particularly striking for pupils in the 9th year. This corresponds with the findings of *E-learning Nordic 2006*, which stresses that in ICT work at school, the pupil often becomes a passive consumer and not an active producer of media content. That leads to a gulf between active, productive use at home and more passive use at school.

An important part of ITU Monitor 2007's work is about operationalizing the concept of digital literacy, in other words, what indicators should be developed to give the theoretical concept an empirical content? How this can be made operational has been solved by emphasizing how pupils use ICT and from this assessing whether these activities make digitally literate users visible. The activities that have been emphasized are: to access, manage, integrate, evaluate and create using ICT (ETS, 2002). The questions that have been asked in the survey are directed towards both the teachers' teaching methods and the pupils' digital practices and can be compared to the information school administrators provide about the school as an organization.

The survey comprised 499 schools, and questions were posed to pupils at the 7th and 9th grades in primary and lower secondary schools and at VK[4] in upper secondary schools. The results for ITU Monitor can give parents, teachers, school owners, and politicians important information about how technology is integrated in teaching and learning in the schools.

ITU Monitor 2007 shows that in several areas a positive development has occurred in the schools' use of ICT over the past couple of years. As far as primary and lower secondary schools are concerned, there has been a particularly notable increase in extent; that is, ICT is used much more frequently for school work by both pupils and teachers than was the case in 2005. Still, there continues to be great variation between primary and lower secondary schools and upper secondary schools—upper secondary schools have come much further than primary and lower secondary schools in integrating the use of ICT in subjects.

The survey shows that many accessible tools and functions are used to only a small extent and that there remains great unutilized potential in the ways ICT is used. The equipment is used only to a limited degree—both with respect to the type of tasks and time spent. There is nonetheless positive development in the use of ICT in the subjects of Norwegian, English and social studies, in the primary and lower secondary schools and in the upper secondary schools. Simultaneously, the difference among pupils in the same grade is still great, and the danger of developing differences in digital skills is ever present.

It is clear that teachers emphasize digital competence to different extents. Great differences exist in the teachers' emphasis on digital literacy. The findings show that interpretation of information is the form of digital competence that teachers emphasize most.

Primary and lower secondary schools are far from a reality where ICT is integrated in all subjects. Even if the teachers use computers more in their work, this increase has primarily been in relation to administrative tasks, not in teaching.

ITU monitor finds three different forms of digital competence among pupils:

- *accessing* information,
- *integrating* information where the information is previously known from before or comes from other sources,
- *creating*, which concerns their digital texts being understandable and, for example, ensuring that illustrations and text fit together.

There are great differences between pupils in the same grade with respect to having and developing digital literacy. ITU Monitor 2007 shows clear differences between pupils in the same grade with respect to what forms of digital skills they have and are developing. It seems as though focusing on mastery (as a learning strategy) and being curious about a subject can have a positive effect on developing knowledge of the use of digital skills. Since there are great differences from pupil to pupil in their focus on mastery and curiosity about subjects, the schools face a great challenge with respect to the pupils who do not learn to use digital tools on their own.

There has been a clear increase in the time spent at computers in the schools between 2005 and 2007. There has been an increase in the use of time spent at computers for school work at home for all grades. There are still great differences between pupils with respect to the extent of their ICT use. The majority of pupils, particularly in primary and lower secondary schools, use computers very little, and we find great variations between pupils in the same grade. The conditions for developing digital competencies are thus very different from school to school. As in 2005, we see that it is simple searches on the internet searches and use of Office programs that dominate.

According to three out of four school administrators, teachers have the basic ICT skills, but they still have some way to go with respect to more educational use of ICT. In addition, the school administrators see a lack of interest among teachers for the educational use of ICT.

When it comes to planning the schools' use of ICT in all areas, it appears that those schools compiling ICT plans for the first time focus most on the operational challenges. Schools that have established ICT plans are more concerned about the educational challenges, such as raising competencies among the staff and pupils. These are conditions that we know are decisive for increasing the professional use of ICT in Norwegian schools.

ITU Monitor shows that there are great differences between what teachers think they are focusing on in their lessons and what the pupils think of their own digital skills. Furthermore, findings show that there are gender differences with respect to having digital skills. In addition, girls report that they are more concerned with learning as much as possible at school than boys are, and this can be characterized as a proactive mastery orientation.

The pupils' background and attitudes to school work are particularly important for the development of digital literacy. Consideration must also be given to the pupils' attitudes to school subjects when lessons are planned and carried out. But the survey also showed that it is a challenge that the teachers have different perceptions of what is key to pupils' digital literacy. Findings indicate that the teachers generally place more emphasis on organizational abilities as a digital skill: pupils should summarize, compare and evaluate information. Managing is only one of several aspects of digital literacy. It is, therefore, necessary to focus more on evaluation, source critique and creative production with digital tools because information searches dominate the use of ICT in teaching. In other words, there is a need to raise consciousness among teachers about what digital literacy is, as well as how they can arrange their teaching in such a way that pupils also develop the skills mentioned above.

The digitally competent school is characterized by its framework, infrastructure, leadership, culture, and educational practice being marked by openness and systematicity.

ITU Monitor 2007 shows that in particular primary and lower secondary schools have some way to go with respect to the utilization of ICT in an open, systematic way, while many upper secondary schools have come much further in this. The schools that have organized ICT efforts with the help of ICT plans that are solidly anchored among the faculty also manage to systematize the work and focus broadly on several decisive measures, such as the development of competencies among teaching staff and the flexible organization of timetables.

Many upper secondary schools, then, have come a long way in the use of ICT in their daily educational work. At the same time there are many upper secondary schools that have not progressed far enough, so that the differences in this area are substantial. Primary and lower secondary schools continue to lag behind upper secondary schools with respect to using ICT as an integrated part of daily school work.

Findings from ITU Monitor show a gap between strategic policy work focused on infrastructure, which is getting continuously better, and ICT in practical pedagogy, which is still lagging behind. Comprehensive implementation of ICT with innovative school development is still inadequate. ICT still remains too much of a "sideline" in national policy. For example, there are major challenges associated with improving ICT focus in teacher training. Existing digital content is not properly utilized and there is a need for the building of digital resources. Last, but not least—the use of ICT pedagogically is lagging behind and still progressing slowly. This shows that building digital competence has a weak position in the Knowledge Promotion Reform.

Digital competence for all is a long-term social project, which requires comprehensive understanding of how to integrate digital tools into schools on a daily basis. It will require adaptability, strategy plans and more resources from central education authorities, school owners and schools.

Digital Stagnation in Teacher Training?

There are a lot of indications that teacher training is out of step with the Knowledge Promotion Reform. Digital competence is not mentioned in the main report from NOKUT's evaluation of teacher training from 2006. Professor Lars Monsen at Lillehammer University College (one of the experts making up the evaluation panel) comments that digital competence was not emphasized in the main report from NOKUT, because working with digital competence was not a priority area at the teacher training colleges.

The gap between the actual investment in digital competence in teacher training and the competence requirements of newly-educated teachers seems to be considerable. NOKUT's partial reports of the assessment exercise also show that there are considerable differences between the institutions. It is necessary to identify, and indeed challenge, the position that digital competence occupies in teacher training. There has not been explicit public investment in this area since *PLUTO* (Project Innovation in Learning, Organisation, and Technology 1999–2003), which led to improved knowledge as a basis for further investment in updating teacher training.

One of the premises for the PLUTO project in teacher training was that in practice, studying and teaching at the educational institutions were modeled on traditional teaching methods rather than developing new ones (Ludvigsen & Rasmussen, 2006). The PLUTO program represented an attempt to change teacher training by focusing on how the students' study and work practices are organized and the relationship between education at the technical colleges and the universities and the practical arena of the school. The overall goal providing the framework for the national strategy to change teacher training was to develop new pedagogical and organizational models for facilitation and implementation of study and learning activities through the use of ICT as an essential tool.

The final PLUTO report, *Modeller på reise (Travelling Models)* (Ludvigsen & Rasmussen 2005) contains concluding statements about the results of the project like: "During the PLUTO program teachers and students at the institutions involved have become high frequency users of various forms of ICT. This was not the case before the program started" (ibid., p. 246). Did the students become better teachers by participating in the PLUTO projects, however?

Several institutions employed examiners who were able to compare the PLUTO students with previous students. The results of the comparison clearly favored the PLUTO students. Furthermore, the failure rate decreased in several subjects. They were given models of working methods that they could use at school like a repertoire of methods, which contributes to increased variation in the pupils' learning and study. They are also exposed to new forms of assessment and examination (ibid.).

There is currently a gap between the requirements for digital competence in the curriculum at all levels and all subjects and the teachers' ability to put the curriculum and the intentions of the Knowledge Promotion Reform into practice. Newly educated teachers have to be digitally competent in order to put the new curriculum into practice, and those educating new teachers also have to master these skills.

The good teacher in the schools of tomorrow is not just an instructor, but a hybrid player who combines academic knowledge and digital *Bildung*. She can see the possibilities inherent in different learning models and varies the use of teaching material and internet usage. The good teacher contributes professionally to different learning situations. She is inspiring, transfers knowledge and she may use multimedia while presenting information. She facilitates a complex computer-simulated experiment and 3D virtual reality games, supervises multi-disciplinary project work, navigates the internet whilst having the ability to critically appraise sources on the internet and comments on the use of computer games in learning situations.

On Top of a Flat World

The objective of digital competence for all and visions of making Norway the leading knowledge nation in the world are ambitious and demanding. Plans and measures by national authorities, school owners and schools are currently insufficiently co-ordinated. We have the knowledge bases and we produce visions and aims, but there is no national or local direction provided and no power to implement these. Both political signals and R&D have maintained for several years that a comprehensive approach to the implementation of ICT *is* a criterion for success. That means that pedagogy, organization and management and technology are considered as a whole, both strategically speaking and in practical terms. Why is this knowledge not applied to a greater extent? When measures are implemented and financed locally and/or nationally, the investment is often only partial and may be unilateral investment in equipment, while competence building for teachers and school managers is not prioritized.

The OECD report *Think Scenarios, Rethinking Education* (2006) from the *Schooling for Tomorrow*[5] program points out that policy making in education is characterized by short-term thinking. In an increasingly complex and unpredictable world, new demands are placed on education and competence. It still seems as if education policy is more preoccupied with the short term and in making the education currently provided more efficient. What are needed are long-term ideas and visions for education in a complex world with continually changing competence requirements. Innovation in schools using ICT is not a one-shot event. It is a long and complex process. According to the OECD, schools cannot be changed from the top down. The change has to include every level of education simultaneously, and there has to be active cooperation between the levels in order to create lasting change. Government, administration, unions and policy-makers are, along with schools, teachers and parents, all players fighting to be heard in such a process of change.

Changing schooling and education is not only a matter of changing the education system but also of innovating the wider socio-economic system, cultural mindsets, and governance frameworks. This is an important observation for understanding the design and revitalization of schooling systems (OECD, 2006, p. 194).

The current generation of decision makers ranging from politicians to teachers sees the world from a different perspective than the digital generation (Green & Hannon, 2007). The young people of today cannot remember a world without the internet, SMS, MSN, iPod, MySpace and Facebook. However, the decision-makers decide how digital media will be used in the schools and the professional world. They make laws and regulations that restrict the potential inherent in digital media. This represents a short-term solution to a long-term challenge. The problem is that schools run the risk of basing their teaching on presentation, communication and assessment methods that are about to become obsolete in both form and content.

Children and young people are increasingly active media users, both as consumers and producers. They are *New Millennium Learners* (Pedro, 2006) according to a recent OECD study. Many pupils develop digital competence at home. They chat about math problems without having been asked to do so by the teacher, they make up fan fiction stories, images and animations, they create music and short movies. Their websites about their final year party van are highly designed, interactive and feature blogs. This productive digital competence can also be applied academically to learning situations at school. Digital competence at school is necessary to educate children and young people for a working life characterized by innovation and value creation. Digital competence is important to the development and continuation of a democratic and inclusive information society.

Norway could become the world's leading nation in digital competence. This requires a long-term comprehensive plan leading up to 2020 with a digital agenda that can help place Norway at the forefront of comparable countries for an education system that provides digital competence, a quality learning outcome and good strategies for learning. A school with a digital focus is inspirational because it meets the pupils on their own terms in their everyday digital world. That's why we need a digital knowledge promotion reform to create the schools of the future.

Developing their digital competence provides children and young people with varied methods for learning, more content resources and a more motivational learning environment. This ultimately adds up to a greater potential for learning more. In the schools of tomorrow pupils will be using digital media with confidence and innovatively to develop skills, knowledge and competencies, which they will need in order to achieve personal goals and become interactive participants in the information society. A digital learning culture entails involvement, ability for critical thought, cooperation and creative problem-solving (Jenkins, 2007). An updated concept of *Bildung* will include consideration of knowledge and identity. This requires fundamental skills, both analogue and digital. Digital *Bildung* is a question of surpassing oneself (Søby, 2001, p. 99). *Bildung* is a continuous investigation of one's own knowledge horizon whilst looking for underlying perspectives and directions. It is a process of seeing oneself and being on top of a flat world.

Endnotes

1. Learning Management System (LMS) is a selection of tools for support of learning activities and their administration. The tools are technically integrated in a common environment and a common database and have therefore shared access to documents, status information and other information.

2. As early as 1983, the National Commission on Excellence in Education in the U.S. launched "technology literacy" as part of the basic education in high school: "a) understand the computer as an information computation and communicating device; b) use the computer in the study of the other basics and for personal and work-related purposes; and c) understand the world of computers, electronics, and related technologies" (National Commission on Excellence in Education, 1983, p. 26).

3. The Educational Testing Service (or ETS) is, according to Wikipedia: "[...] the world's largest private educational testing and measurement organization, operating on an annual budget of approximately $900 million. ETS develops various standardized examinations primarily in the United States, but they also administer tests. Many of the assessments they develop are associated with entry to U.S. (undergraduate) and (graduate) institutions."

4. The second year of upper secondary education.
5. In the program "Schooling for Tomorrow," the OECD has compiled a number of reports about trends and scenarios for schools. The six scenarios are:
   a. Bureaucratic School Systems Continue
   b. Schools as Focused Learning Organisations
   c. Schools as Core Social Centres
   d. Extending the Market Model
   e. Learning Networks and the Network Society
   f. Teacher Exodus and System Meltdown

References

Arnseth, H.C., O. Hatlevik, V. Kløvstad, T. Kristiansen, G. Ottestad. (2007). *ITU monitor 2007.* Oslo: Universitetsforlaget.

Bateson, G. (1979). *Mind and nature.* London: Bantam Books.

Buckingham, D. (2006). Defining digital literacy. *Digital Kompetanse, 1(4),* 263–277.

Castells, M. (1996). *Information age: Economy, society and culture. The Rise of the Network Society.* Oxford: Blackwell.

———. (2001). *The Internet galaxy: Reflections on the Internet, business, and society.* New York: Oxford University Press.

Castells, M , & M. Ince. (2003). *Conversations with Manel Castells.* Cambridge. Polity Press.

Connolly, W. E. (1993). *The terms of political discourse.* Oxford: Blackwell.

Dale, E. L. (1972). *Pedagogikk og samfunnsforandring.* Oslo: Gyldendal.

Educational Testing Service. (2002). *Digital transformation: A framework for ICT literacy.* Retrieved November 25, 2007 from http://www.ets.org/research/ictliteracy/ictreport.pdf

Erstad, O. (2005). *Digital kompetanse i skolen—en innføring.* Oslo: Universitetsforlaget.

Erstad, O., V. Kløvstad, Kristiansen, T. & Søby, M. (2005). *ITU Monitor 2005. På vei mot digital kompetanse i grunnopplæringen.* Oslo: Universitetsforlaget.

European Commission. (2003). *eLearning: Better eLearning for Europe.* Directorate-General for Education and Culture. Luxembourg. Office for Official Publications of the European Communities.

———. (2004): *Key competences for lifelong learning: a European reference framework.* Directorate-General for Education and Culture. Retrieved November 22, 2007 from http://europa.eu.int/comm/education/policies/2010/doc/basicframe.pdf

European Parliament and the Council of the European Union. (2006). Retrieved December 2, 2007, from http://eurex.europa.eu/LexUriServ/site/en/oj/2006/l_394/l_39420061230en00100018.pdf

FAFO. (2007). *Analyse av aktivitetsrapportering 2006 -kompetanse for utvikling.* Retrieved November 27, 2007: http://www.utdanningsdirektoratet.no/upload/Kompetanseutvikling/Aktivitetsrapportering_2006%20_Fafo.pdf

Gilster, P. (1997). *Digital Literacy.* New York: Wiley.

Green, H. & Hannon, C.. (2007). *Their Space. Education for a digital generation.* London: Demos.

Hellesenes, J. (1975). *Sosialisering og teknokrati.* Oslo: Gyldendal.

Hermann, S. (2005). Kompetencebegrebets udviklingshistorie. *Tidsskrift for læreruddannelse og skole, 71,* 7–17.

ITU (2006). *Digital skole hver dag.* www.itu.no/digital_kompetanse/1130232549.62

———— (2007a). "Sammensatte tekster." *Forskning viser no. 8,* Oslo: ITU.

———— (2007b). "Kildekritikk og kritisk refleksjon." *Forskning viser no. 10,* Oslo: ITU.

Jenkins, H. (2007). Confronting the challenges of participatory culture: Media education for the 21st century. *Digital Kompetanse, 2(1),* 23–34.

Kirke-utdannings-og forskningsdepartementet. (1996). *IT i norsk utdanning. Plan for 1996 –1999.* Oslo: KUF.

————. (2000). *IKT i norsk utdanning. Plan for 2000–2003.* Oslo: KUF.

Koschmann, T. (Ed.). (1996). *CSCL: Theory and practice of an emerging paradigm.* Mahwah, New Jersey: Lawrence Erlbaum.

Krumsvik, R. J. (Ed.). (2007). *Skulen og den digital læringsrevolusjonen.* Oslo: Universitetsforlaget.

Kunnskapsdepartementet. (2006). *Læreplanverket for kunnskapsløftet.* Midlertidig utgave juni 2006. Oslo: Utdanningsdirektoratet.

Ludvigsen, S., & Rasmussen, I. (2006). "Modeller på reise," *Digital Kompetanse 1(3),* 227–251.

Liestøl, G. (2006). Sammensatte tekster—sammensatt kompetanse. *Digital Kompetanse.1(4),* 277–306.

Ludvigsen, S. & Rasmussen, I. (2005). Interdependency in teaching and learning: A study of project work and ICT. Paper presented: ISCAR Spain, Seville. September.

McLuhan, M. (1968). *Mennesket og media.* Oslo: Gyldendal.

Metcalfe, L. (1993): Public management: From imitation to innovation. In Kooimann, J. (Ed.), *Modern governance: New government-society interactions.* London: Sage Publications.

National Commission on Excellence in Education (1983). *A nation at risk: The imperative for educational reform.* Washington, DC: U.S. Department of Education.

MOD. (2005). *eNorge 2009: det digitale spranget.* Oslo: Moderniseringsdepartementet.

Nisbet, R. (1969). *Social change and history: Aspects of the Western theory of development.* New York: Oxford University Press.

NOKUT (2006) *Evaluering av allmennlærerutdanningen i Norge 2006. Del 1: Hovedrapport.* Oslo: NOKUT.

NOKUT. (2006). *Evaluering av allmennlærerutdanningen i Norge 2006. Del 2: Institusjonsrapporter.* Oslo: NOKUT.

Norwegian Directorate for Education and Training (2006). *Læreplanverket for Kunnskapsløftet.* (Midlertidig utgave, juni 2006) Oslo.

—————— (2007). *The State of Digital Comptenece.* Oslo: Norwegian Ministry of Education and Research.

NOU 2003:16 *I første rekke. Forsterket kvalitet i en grunnopplæring for alle.* (Kvalitetsutvalget).

OECD (2002): *The definition and selection of key competencies.* Retrieved December 5, 2007 from http://www.oecd.org/dataoecd/47/61/35070367.pdf

———— (2006). *Schooling of tomorrow: Think scenarios, rethink education.* Paris: OECD Publishing.

Ong, W. J. (1982). *Orality and literacy: The technologizing of the word.* London: Methuen.

Pedro, F. (2006). *New millennium learners*. Retrieved December 5, 2007 from http://www.oecd.org/dataoecd/1/1/38358359.pdf

Platon (1962). *Elskoven og sjelen. Faidros.* Oslo: Aschehoug.

Rambøll Management. (2006). *E-learning Nordic 2006.* Glostrup: Rambøl.

Rousseau, Jean-Jacques. (1962). *Emile eller om opdragelsen.* Ringkjøping: Borgens forlag.

Rychen D.S. & Salganik L.H. (Eds.). (2003). *Key competencies for a successful life and a well-functioning society.* Göttingen: Hogrefe & Huber Publishers.

Snow, C. (1959). *The two cultures and the scientific revolution.* Cambridge: Cambridge University Press.

Søby, M. (1998). Vi er alle kyborgere. *Nordisk Pedagogik, 18(1),*16–36.

———. (2001). Interaktivitet—moteord eller passord til fremtidens pedagogikk? *Utbildning & Demokrati, 10(1),* 85–103.

———. (2003). *Digital kompetanse: Fra 4. basisferdighet til digital dannelse. Et problemnotat.* Oslo: ITU.

Tyner, K. (1998). *Literacy in a digital world.* Mahwah, New Jersey: Lawrence Erlbaum.

Utdanningsdirektoratet. (2005). *Kartlegging og rapportering av utstyrs- og driftssituasjonen grunnopplæringen.* Oslo.

———. (2006). *Læreplanverket for Kunnskapsløftet.* (Midlertidig utgave, juni 2006) Oslo.

——— (2007a). *Et digitalt kompetanseløft for alle? En midtveisrapport for Program for digital kompetanse 2004–2008.* Oslo.

——— (2007b). *Utstyrs-og driftssituasjonen i grunnopplæringen 2006–2007.* Oslo.

Utdannings-og forskningsdepartementet (2004a). *Program for digital kompetanse 2004–2008.* Oslo.

——— (2004b): *Kompetanse for utvikling. Strategi for kompetanseutvikling i grunnopplæringen 2005–2008.*

White paper no. 30. (2003–2004). *Kultur for læring.* Oslo: Utdannings-og forskningsdepartementet

White paper no. 17. (2006–2007). *Eit informasjonssamfunn for alle.* Oslo: Fornyings-og administrasjonsdepartementet.

CHAPTER SEVEN

Digital Literacy and the "Digital Society"

ALLAN MARTIN

Society and the Digital

Throughout most of Europe and many other parts of the world, we live today in a society permeated by the digital, where our actions are frequently mediated by digital tools, and the objects we encounter are frequently shaped by digital intervention. The mobile phone and the MP3 player are the most visible personal artifacts of this society, whilst the PC is the ubiquitous gateway to cyber-activity, at work and at home.

Yet it would be wrong to think that we live in "The Digital Society," for this suggests that society is *made* by the digital, and that its essential characteristics have been created because of the development of digital technology. Many of the superlatives coined over the last forty years to characterize the impact of the computer, including the "Electronic Revolution" (Handel, 1967), the "Technetronic Age" (Brzezinski, 1970), the "Microelectronics Revolution" (Large, 1980; Forrester, 1980), the "Computer Age" (Dertouzos & Moses, 1979), take a feature of social activity—namely, its visible technology—and imply that this feature is its essence.

In the 1990s these terms were superseded by the "Information Society" and the "Knowledge Revolution." According to a House of Lords Select Committee report (House of Lords, 1996: §1.6), "The world is undergoing a technological revolution and entering the age of the Information Society." In 1995 Bill Gates claimed that "The information revolution is just beginning." (Gates, 1995, p. 21) The notion of the "Information Society" is now a commonplace in documents produced by the UK government and by European agencies. A European Commission report *Europe and the Global Information Society* stated that, "throughout the world, information and communications technologies are bringing about a new industrial revolution which already looks to be as important and radical as those which preceded it." (European Commission, 1994, p. 4) The UK government's 1998 Green Paper *The Learning Age: A Renaissance for a New Britain* announced in its very first sentence that "We are in a new age—the age of information and of global competition." (DfEE, 1998, p. 9). The implication is that this new social form has been created because of technological change.

The reality is less simple. A major problem with notions like the "technological revolution" and the "information society" is that they are powerful metaphors with a misleading message. The message is misleading in three ways.

First, these terms create the impression that social change is determined by technology. This reification of a human product obscures the fact that change and, indeed, technology are both products of human action and interaction, and that the relationship of technology to social change is a non-simple one. Even the most spectacular inventions are rooted in a social order that enabled them to happen and then identified them as important. For their own purposes people, mainly at the behest of governments and business, have striven to make and to better this technology. Billions of dollars have been invested to make digital technology more powerful, more versatile, more cost-effective, more profitable. We have made the "Information Society" and the "Digital Age" for ourselves.

Second, the attribution of events to a technological origin is also a moral statement, since the blaming of human actions on technology allows humans to escape responsibility for actions which were the results of their own choices.

Third, ideas like "technological revolution" and "information society" suggest that social change is characterized by revolutions, i.e., sudden, unexpected, and simple shifts from one mode of activity to another; whereas in reality change displays more embeddedness in what came before, and all inventions have an ancestry. Whilst these terms, and others like them, do capture a very prominent feature of contemporary society, and one that does point to changes in behavior, they mask the essential continuity of the social, economic and political order. We continue to live in a hierarchical and unequal society dominated by the ideology of free market capitalism, and the "digital divide" merely adds another dimension to inequalities which have already long existed.

Society, however, is not static. The "end of history" has proved, even to those who believed it, a false dusk. Zygmunt Bauman (2000, 2001), among others (e.g., Beck, 1992; Giddens, 1990, 1999), identifies the gradual dissolution of the classical industrial order into a society of uncertainty and risk, in which nothing can be predicted and the long term becomes meaningless. Heavy industry, the nation state, and institutionalized religion, the three pillars of the "modern" order are dissolving, robbing individuals of the certainties they once provided: of work, order and belief. In the era of what Bauman calls "liquid modernity" individuals fall back upon short-term satisfaction, temporary goals, and the acquisition of objects or consumption of services. Everything is short term because long-term certainties cannot be guaranteed.

Society is being transformed by the passage from the "solid" to the "liquid" phases of modernity, in which all social forms melt faster than new ones can be cast. They are not given enough time to solidify and cannot serve as the frame of reference for human actions and long-term life-strategies because their allegedly short life expectation undermines efforts to develop a strategy that would require the consistent fulfillment of a "life-project." (Bauman, 2005, p. 303)

For those who do not belong to the global elite, life has become an individual struggle for meaning and livelihood in a world that has lost its predictability—what Ulrich Beck calls "The Risk Society" (Beck, 1992). Consumption has become the only reality, the main topic of TV and of conversation, and the focus of leisure activity. The modes of consumption become badges of order, so that to wear a football strip of a certain team (themselves now multinational concerns) or a logo of a multinational company become temporary guarantors of safety and normality.

In this society, the construction of individual identity has become the fundamental social act. The taken-for-granted structures of modern (i.e., industrial) society—the nation state, institutionalized religion, social class—have become weaker and fuzzier as providers of meaning and, to that extent, of predictability. Even the family has become more atomized and short term. Under such conditions individual identity becomes the major life-project. You have to choose the pieces (from those available to you) rather than having them (largely) chosen for you. In this context, awareness of the self assumes new importance: reflexivity is a condition of life; a life that needs to be constantly active and constantly re-created. And care is needed, because each individual is responsible for their own biography. Risk and uncertainty have become endemic features of the personal biography, and individual risk-management action is thus an essential element of social action (Beck, 1992, 2001). The community can be no longer regarded as a given that confers aspects of identity, and the building of involvement in communities has become a conscious action-forming part of the construction of individual identity. Individualization has positive as well as negative aspects: the freedom to make one's own biography has never been greater, a theme frequently repeated in the media. But the structures of society continue to distribute the choices available very unequally, and the price of failure is greater since social support is now offered only equivocally.

One element of continuity is the free-market economy, validated by the collapse of the communist alternative. But in a globalized society the free-market economy has become transformed into a supranational order in which the elite, no longer loyal or beholden to any one country, and itself highly mobile in terms of its location and lifestyle, deploys capital on a global basis, moving resources from state to state, from continent to continent, responding continually to changes in commodity prices, raw material availability and transport and labor costs. In this context the role of the state as provider, and occasionally as enforcer, has become less relevant to the capitalist order. Accordingly, the state has begun to step back from these roles, reducing its activity or passing it over to the private sector. Although it has not created it, digital technology is nonetheless complicit in the enablement of a global society, and has become essential to the accomplishment of most official and commercial activities, and many personal ones too. The digital, which was initially a tool to achieve faster and more efficiently activities we already performed, has enabled activities previously considered unimaginable, including globalization itself.

The causes of this direction of social change are many and, as with all social changes, technology is simultaneously its tool, its medium and its reflection. Digital technology is thus both means and symptom of social change. Digital technologies have enabled the globalization of business processes and of commercial cultural output, and also the surveillance of individuals, and the capture of individual identities through data mining and collation. The digital is well implicated in the genesis and maintenance of this "post-modern" society, but it is the major actors in that society who have driven it so, directing research and investment in "new technology" in order to reap substantial profits or to sustain hegemonic economic and political ambitions. Smaller players have also contributed by providing taxes for state investment and queuing to purchase the latest in digital equipment and to consume the latest product of the digital media. For the ordinary individual (one who is not a member of international economic, political and media élites), the choices may seem very limited—to be part of the consumer society has become a vital source of meaning and identity.

The digital is (almost) ubiquitous, and its possibilities are both creative and destructive in the quest for identity. Digital tools enable the individual to present him/herself to the rest of society by creating and broadcasting statements (developing blogs or personal websites, contributing to online fora, sending email, texting, presenting a curriculum vitae, etc.) or multi-media objects (mounted on social collection sites). They also enable social identity development, making oneself in interaction with others, members of "strong" groups such as family or friends, or "weak" groups such as online "communities." Yet the individual is also threatened by external drivers pressing templates of identity upon him/her. Images of "normality" in the media are presented via digital as well as non-digital means. Data collection by external agencies is everywhere, on overtly digital actions (such as using search engines or buying online) as well as those in which the digital aspect is masked (such as buying clothes, attending a concert, flying from one place to another or staying at hotels). In most cases the data collection and analysis are covert, and the fate of the data is unknown to those from whom it derives. But its digital nature gives it for those who use it an element of certainty that can aggressively confront the uncertainties of real life. Digital data representations can be regarded as more real than our own versions of identity. Online shops now habitually tell me what I want to buy before I've had a chance to express a preference—they know me better than I know myself. Bigger dangers also can present themselves, especially that of identity theft, perhaps the ultimate digital crime.

The challenge for individuals in the dissolving social order of late modernity is to maintain, or regain, some control of their own destinies, to retain an involvement in the creation of meaning. Hence the centrality of notions of literacy which, despite the differences in the ways they are framed, all partake of the individual's engagement with the meanings current in society. The idea of literacy expresses one of the fundamental characteristics of participation in society, and the widening application of the word has seen it used to characterize all of the necessary attributes of social being.

Claire Bélisle (2006) characterizes the evolution of literacy concepts in terms of three models. The *functional model* views literacy as the mastery of simple cognitive and practical skills and ranges from the simple view of literacy as the mechanical skills of reading and writing to a more developed approach (evinced by, e.g., UNESCO, 2006) regarding literacy as the skills required for functioning effectively within the community. The *socio-cultural practice* model takes as its basis that the concept of literacy is only meaningful in terms of its social context and that to be literate is to have access to cultural, economic and political structures of society; in this sense, as Brian Street (1984) has argued, literacy is ideological. The *intellectual empowerment* model argues that:

> Literacy not only provides means and skills to deal with written texts and numbers within specific cultural and ideological contexts, but it brings a profound enrichment and eventually entails a transformation of human thinking capacities. This intellectual empowerment happens whenever mankind endows itself with new cognitive tools, such as writing, or with new technical instruments, such as those that digital technology has made possible. (Bélisle, 2006, pp. 54–55)

In viewing literacy within the context of a digitally infused society as, at one level functional, at another socially engaged, and at a third as transformative, we can see it as a powerful tool for the individual and the group to understand their own relationship to the digital: to be aware of the role of the digital in their own development and to control it; to place the digital at the disposal of their own goals and visions. Gaining a literacy of the digital is thus one means by which the individual can retain a hold on the shape of his/her life in an era of increasing uncertainty.

Literacies of the Digital

We can identify several "literacies of the digital," mostly originating in the predigital period but presented as routes to understanding phenomena which have become more significant or even transformed in digital contexts.

Computer, IT or ICT Literacy

This has been identified as a need from the late 1960s, when it became clear that access to computers could be enjoyed by large numbers. We can see concepts of computer literacy as passing through three phases: the *Mastery* phase (up to the mid-1980s), the *Application* phase (mid-1980s to late-1990s) and the *Reflective* phase (late-1990s on) (Martin, 2003).

In the *Mastery* phase the computer is perceived as arcane and powerful, and emphasis is placed on gaining specialist knowledge and skill to master it. "Computer Basics," whatever they may be called, consist of how the computer works (simple computer science), and how to program it (using whatever languages were current at the time), sometimes with additional input on the "social and economic effects" of computers.

The *Application* phase began towards the end of the 1980s with the appearance of simple graphical user interfaces and easy-to-use mass market applications, which opened up computers to mass usage. In this phase the computer is seen as a tool that can be applied in education, work, leisure and the home. How to use applications software becomes the focus of literacy activity, and definitions of computer or IT literacy focus on practical competences rather than specialist knowledge. This is accompanied by the appearance of mass certification schemes focusing on basic levels of IT competence.

Movement to the *Reflective* phase was stimulated by realizations that IT could be a vehicle through which student-centered pedagogies, championed by innovators since the 1960s, could at last be realized. There is an awareness of the need for more critical, evaluative and reflective approaches to using IT. At the reflective level specific skills are superseded by generic skills or meta-skills, as evident in the definition formulated by the OECD-ILO PISA project:

> ICT literacy is the interest, attitude and ability of individuals to appropriately use digital technology and communication tools to access, manage, integrate and evaluate information, construct new knowledge, and communicate with others in order to participate effectively in society. (van Joolingen, 2004)

The term "fluency" was proposed by a U.S. National Research Council report (NRC, 1999) to suggest the greater intellectual challenge proposed. The report comments that

> Generally, 'computer literacy' has acquired a 'skills' connotation, implying competency with a few of today's computer applications, such as word processing and e-mail. Literacy is too modest a goal in the presence of rapid change, because it lacks the necessary 'staying power.' As the technology changes by leaps and bounds, existing skills become antiquated and there is no migration path to new skills. A better solution is for the individual to plan to adapt to changes in the technology. (NRC, 1999, p. 2)

The U.S. Educational Testing Service report (ETS, 2002) takes a clear position on the reflective nature of ICT literacy:

> ICT literacy cannot be defined primarily as the mastery of technical skills. The panel concludes that the concept of ICT literacy should be broadened to include both critical cognitive skills as well as the application of technical skills and knowledge. These cognitive skills include general literacy, such as reading and numeracy, as well as critical thinking and problem solving. Without such skills, the panel believes that true ICT literacy cannot be attained. (ETS, 2002, p. 1)

The report defines ICT literacy as follows:

> ICT literacy is using digital technology, communications tools, and/or networks to access, manage, integrate, and create information in order to function in a knowledge society. (ibid., p. 2) . . .
> The five components represent a continuum of skills and knowledge and are presented in a sequence suggesting increasing cognitive complexity. . . .
> > **Access**—knowing about and knowing how to collect and/or retrieve information.
> > **Manage**—applying an existing organizational or classification scheme.
> > **Integrate**—interpreting and representing information. It involves summarizing, comparing and contrasting.
> > **Evaluate**—making judgments about the quality, relevance, usefulness, or efficiency of information.
> > **Create**—generating information by adapting, applying, designing, inventing, or authoring information. (ibid., p. 3)

It is possible that this three-phase development of ICT literacy, from skills through usage to reflection, is paralleled in the evolution of the other literacies considered here. We should note that the earlier phases remain as subordinate layers, so that literacy concepts become more complex and multi-layered as they develop.

Technological Literacy

The idea of technological literacy emerged in the 1970s as a response to two very different concerns: the growing awareness of the enormous potential danger of technological developments for the environment and for humanity and the growing fear that ignorance of developing technologies would render the workforce in countries like the U.S. and Britain vulnerable to competition from countries with more technological awareness (Waks, 2006). The result was an uneasy marriage of the two concerns, since one favored a skills-based vocational approach (with a preference for a behaviorist pedagogy) and the other a critical, action-oriented "academic" approach (with a liking for a more constructivist pedagogy) (Dakers, 2006b). This compromise is reflected in the *Technology for All Americans* materials, funded by the U.S. government:

> Technological literacy is the ability to use, manage, and understand technology:
> - The ability to use technology involves the successful operation of the key systems of the time. This includes knowing the components of existing macrosystems, or human adaptive systems, and how the systems behave.
> - The ability to manage technology involves insuring that all technological activities are efficient and appropriate.
> - Understanding technology involves more than facts and information but also the ability to synthesize the information into new insights. (ITEA, 1996, p. 5)

A major criticism of these developments is that, despite the rhetoric, the critical element of technological literacy is insufficiently developed or implemented, and it must engage the industrial application of technology with deeper understanding of the social and political involvement of technology (Michael, 2006). This will involve more reference to theorists like Feenberg (1999) who critically address the role of technology in society.

Information Literacy

This developed in the U.S. since the late 1980s as a re-focusing of "bibliographic instruction" in academic libraries, in the light of the trend towards student-centered learning, and thus arose in a largely pre-digital context. With the increasing perception of the Worldwide Web as a seemingly infinite source of information, the information literacy movement gained more urgency. The U.S. Association of College and Research Libraries, focusing on higher education, presents a set of performance indicators based on five "standards":

> The information literate student:
> i. determines the nature and extent of the information needed;
> ii. accesses needed information effectively and efficiently;
> iii. evaluates information and its sources critically and incorporates selected information into his or her knowledge base and value system;
> iv. uses information effectively to accomplish a specific purpose;
> v. understands many of the economic, legal, and social issues surrounding the use of information and accesses and uses information ethically and legally. (ACRL, 2000, pp. 8–13, passim)

Information literacy has influenced librarians on a worldwide basis (see

Rader, 2003), and is seen as important by national and international bodies. An "Information Literacy Meeting of Experts," held in Prague in 2003, led to the so-called "Prague Declaration" (UNESCO, 2003) stressing the global importance of information literacy in the context of the "Information Society." It includes the statement that:

> Information Literacy encompasses knowledge of one's information concerns and needs, and the ability to identify, locate, evaluate, organize and effectively create, use and communicate information to address issues or problems at hand; it is a prerequisite for participating effectively in the Information Society, and is part of the basic human right of life long learning. (ibid., p. 1)

What emerges from the report of the meeting is that information literacy is not simply about digital information, that in fact there is a much wider challenge of which digital environments form only one part. This is a welcome counterbalance to the assumption, easily made in the developed world that information literacy is only, or mainly, about digital information. Research by Bill Johnston and Sheila Webber suggests that digital factors have less impact on academics' perceptions of information literacy than do their pedagogical approaches. Johnston and Webber (2003) underline the media-independent nature of information literacy with their own definition:

> the adoption of appropriate information behaviour to obtain, through whatever channel or medium, information well fitted to information needs, together with critical awareness of the importance of wise and ethical use of information in society. (http://dis.shef.ac.uk/literacy/project/about.html)

Media Literacy

Also known as "media education," media literacy has developed from the critical evaluation of mass media and is a major educational and research activity in both the U.S. and Europe. Tyner (1998, p. 113) defines it as follows:

> Media literacy attempts to consolidate strands from the communication multiliteracies that correspond with the convergence of text, sound and image, including the moving image. It has been associated with the ability to make sense of all media and genre, from the more classic educational fare to popular culture.

The Alliance for a Media Literate America offers an alternative definition on its website:

Within North America, media literacy is seen to consist of a series of communication competencies, including the ability to ACCESS, ANALYZE, EVALUATE and COMMUNICATE information in a variety of forms including print and non-print messages. Interdisciplinary by nature, media literacy represents a necessary, inevitable and realistic response to the complex, ever-changing electronic environment and communication cornucopia that surrounds us. (http://www.amlainfo.org/home/media-literacy)

Focusing on work in schools, Hobbs (1998) proposes a new definition of literacy based on the ideas of the media literacy movement:

Literacy is the ability to access, analyze, evaluate and communicate messages in a variety of forms. Embedded in this definition [are] both a process for learning and an expansion of the concept of "text" to include messages of all sorts. This view of literacy posits the student as being actively engaged in the process of analyzing and creating messages and as a result, this definition reflects some basic principles of school reform which generally include:

- inquiry based education
- student centered learning
- problem solving in cooperative teams
- alternatives to standardized testing
- integrated curriculum (Hobbs, 1998, p. 8)

There is much similarity between definitions of media literacy and information literacy, which suggests that the generic competences are very similar. Media literacy is focused more on the nature of various genres of medium and the way in which messages are constructed and interpreted—in this perspective the characteristics of the author/sender and the receiver are crucial in understanding the meaning of the message and its content. Information literacy has tended to focus on the ways in which information is accessed and the evaluation of the content.

Visual Literacy

Visual literacy has developed out of art criticism and art education and was initially concerned with perception and the way in which artists and designers have used perspective, ratio, light, color and other techniques of visual communication. The term was coined in 1969 by John Debes, founder of the International Visual Literacy Association (IVLA):

Visual Literacy refers to a group of vision-competencies a human being can develop by seeing and at the same time having and integrating other sensory experiences. The development of these competencies is fundamental to normal human learning. When developed, they enable a visually literate person to discriminate and interpret the visible actions, objects, symbols, natural or man-made, that he encounters in his environment. Through the creative use of these competencies, he is able to communicate with others. Through the appreciative use of these competencies, he is able to comprehend and enjoy the masterworks of visual communication. (http://www.ivla. org/org_what_vis_lit.htm)

Wilde and Wilde (1991, p. 12) link visual problem-solving to "the quest for visual literacy" and offer this as "the best hope for creating future generations of visually literate designers." Dondis, however, emphasizes that this approach can enable everybody (not merely the artistic elite) to engage with the visual aspects of culture and thus sees visual literacy as very much paralleling classical literacy:

Literacy means that a group shares the assigned meaning of a common body of information. Visual literacy must operate somewhat within the same boundaries. . . . Its purposes are the same as those that motivated the development of written language: to construct a basic system for learning, recognizing, making, and understanding visual messages that are negotiable by all people, not just those specially trained, like the designer, the artist, the craftsman, and the aesthetician. (Dondis, 1973, p. x)

Visual images have always been a powerful medium for the interpretation of information and the communication of meaning, in science as well as art, and in dealing with the exigencies of everyday life. The wealth and complexity of visual imagery which is possible using digital tools emphasize the power of the visual. The website visualcomplexity.com, for instance, offers many examples of how visual structures are used in the processes of interpreting data and creating new knowledge.

Communication Literacy

This underlines the importance of communication as a human activity—indeed, as a basis of social interaction—and is seen as a basic personal attribute, whether mediated orally or digitally. But the advent of the digital, offering instant communication to one or many disassembled from a face-to-face situation requires the user to be more aware of the nature and implications of the medium. The website of the Winnipeg School Division defines communication literacy in the following terms:

Learners must be able to communicate effectively as individuals and work collaboratively in groups, using publishing technologies (word processor, database, spreadsheet, drawing tools . . .), the Internet, as well as other electronic and telecommunication tools. (http://www.wsd1.org/techcont/introduction.htm)

Meanwhile, a course proposal at the University of Washington offers a broader definition of communication literacy:

These introductory courses will focus on identifying communication as a unique area of study in that its content (communication processes) is mirrored in its use. Indeed, several communication theorists discuss "practical theory" or "communication praxis" to identify the nature of what we study as both a theoretical and pragmatic endeavor. . . . This will include, but not be limited to, increased understanding and effectiveness in public speaking, non-fiction writing, media viewing and reading, new media and technology, as well as cultural, intercultural, and international interaction. (http://www. artsci.washington.edu/services/Curriculum/2001Awards/Communication.pdf)

Literacy theorists have also recognized the significance of the digital in shaping the contexts within which literacy is to be understood. Lankshear and Knobel (2003, pp. 16–17) describe "new literacies" with reference to the digital:

The category of 'new literacies' largely covers what are often referred to as 'post-typographic' forms of textual practice. These include using and constructing hyperlinks between documents and/or images, sounds, movies, semiotic languages (such as . . . emoticons ('smileys') used in email, online chat space or in instant messaging), manipulating a mouse to move around *within* a text, reading file extension and identifying what software will 'read' each file, producing 'non-linear' texts, navigating three-dimensional worlds online and so on.

To these new literacies, which do not all necessarily involve ICT, they add further literacies which are new in a chronological sense or in being considered as literacies; these include: scenario planning, zines, multimediating, e-zining, meme-ing, blogging, map rapping, culture jamming, and communication guerrilla actions (ibid., pp. 23–49).

Focusing on the idea of a range of distinct but interrelated literacies, some commentators use the plural terms "literacies," "multiple literacies" or "multiliteracies." Kellner (2002, p. 163) prefers the term "multiple literacies" which "points to the many different kinds of literacies needed to access, interpret, criticize, and participate in the emergent new forms of culture and society," but also refers to "technoliteracies" (Kahn & Kellner, 2006). Snyder calls her 2002 book *Silicon Literacies*, although in the text itself tends to refer to "literacy practices." Kress (2003), however, resists the notion of a multiplicity of literacies, suggesting that it leads to serious conceptual confusion, and argues that it is instead necessary to develop a new theoretical framework for *literacy* which can use a single set of concepts to address its various aspects. Tyner (1998, pp. 63–68) recognizes the need to refer to multiliteracies but prefers to identify groups of linked literacies while retaining "literacy" as an overarching concept.

It is clear that there is considerable overlap between the literacies outlined above. In some cases, the definitions of the different literacies are almost identical and only nuanced in different directions, as a result of their pathways from pre-digital foci and their sense of the concerns of the particular community they have developed to serve. Part of the convergence also involves the evolution of literacies from a skills focus through an applications focus towards a concern with critique, reflection and judgment and the identification of generic cognitive abilities or processes, or meta-skills. In this way the digital literacies define themselves as being concerned with the application of similar critical/reflective abilities in slightly different fields of activity. Alongside this has been an identification of student-centered pedagogy as the appropriate vehicle for literacy activities.

Digital Literacy

Is it possible then to talk of a "digital literacy"? This term was popularized by Paul Gilster, who, in his book of the same name, defined it as:

> the ability to understand and use information in multiple formats from a wide range of sources when it is presented via computers. The concept of literacy goes beyond simply being able to read; it has always meant the ability to read with meaning, and to understand. It is the fundamental act of cognition. Digital literacy likewise extends the boundaries of definition. It is cognition of what you see on the computer screen when you use the networked medium. It places demands upon you that were always present, though less visible, in the analog media of newspaper and TV. At the same time, it conjures up a new set of challenges that require you to approach networked computers without preconceptions. Not only must you acquire the skill of finding things, you must also acquire the ability to use these things in your life. (Gilster, 1997, pp. 1–2)

Gilster identifies critical thinking rather than technical competence as the core skill of digital literacy and emphasizes the critical evaluation of what is found on the Web, rather than the technical skills required to access it. He also emphasizes, in the last sentence, the relevant usage of skills "in your life," that digital literacy is more than skills or competences.

The Canadian SchoolNet National Advisory Board (SNAB) takes a similar approach, focusing not only on the mastery of skills but the ability to use them in appropriate circumstances.

> Digital literacy presupposes an understanding of technical tools, but concerns primarily the capacity to employ those tools effectively. Hence, digital literacy begins with the ability to retrieve, manage, share and create information and knowledge, but is consummated through the acquisition of enhanced skills in problem solving, critical thinking, communication and collaboration. (SNAB, 2001, p. 3)

The SNAB links the need for digital literacy to the importance of developing innovative capacities.

> In an interactive, connected world, Canada's ability to foster innovation is linked to its ability to develop a critical mass of knowledge workers and digitally literate citizens— i.e. based on the ability to use information and communications technologies (ICT). (ibid., p. 1)

These sentiments are echoed in the report *Digital Horizons* of the New Zealand Ministry of Education.

> Digital literacy is the ability to appreciate the potential of ICT to support innovation in industrial, business and creative processes. Learners need to gain the confidence, skills, and discrimination to adopt ICT in appropriate ways. Digital literacy is seen as a 'life skill' in the same way as literacy and numeracy. (Ministry of Education, 2003, p. 5)

The European Commission which, in the last two years has adopted digital literacy as a key concept, leaves the definition vague, speaking in terms of "the ability to use ICT and the Internet" (European Commission, 2003, p. 3) or "the ability to effectively use ICT" (ibid., p. 14).

In an earlier paper I offered "eLiteracy" as a synthesizing concept, defining it as:

> the awarenesses, skills, understandings, and reflective-evaluative approaches that are necessary for an individual to operate comfortably in information-rich and ICT-supported environments. An individual is eLiterate to the extent that they have acquired these awarenesses, skills, and approaches. . . .
> For the individual, eLiteracy consists of:
> a. awareness of the ICT and information environment;
> b. confidence in using generic ICT and information tools;
> c. evaluation of information-handling operations and products;
> d. reflection on one's own eLiteracy development;

e. adaptability and willingness to meet eLiteracy challenges. (Martin, 2003, p. 18)

This definition seems to fit well with views of digital literacy.

A further perspective is provided by Søby who, in a report for the Norwegian Ministry of Research and Education, draws attention to the concept of "Digital Bildung":

> Digital *bildung* expresses a more holistic understanding of how children and youths learn and develop their identity. In addition, the concept encompasses and combines the way in which skills, qualifications, and knowledge are used. As such, digital *bildung* suggests an integrated, holistic approach that enables reflection on the effects that ICT has on different aspects of human development: communicative competence, critical thinking skills, and enculturation processes, among others. (Søby, 2003, p. 8; see also Søby, Ch. 6, this volume)

Søby uses the German term *Bildung* to suggest the integrated development of the individual as a whole person. The process of *Bildung* goes on throughout life, affects all aspects of the individual's thought and activity, and affects understandings, interpretations, beliefs, attitudes and emotions as well as actions. It represents the making of the individual both as a unique individual and as a member of a culture.

On the basis of the discussion above, *digital literacy* can be seen as including several key elements:

i. Digital literacy involves being able to carry out successful digital actions embedded within work, learning, leisure, and other aspects of everyday life;

ii. Digital literacy, for the individual, will therefore vary according to his/her particular life situation and also be an ongoing lifelong process developing as the individual's life situation evolves;

iii. Digital literacy is broader than ICT literacy and will include elements drawn from several related "digital literacies";

iv. Digital literacy involves acquiring and using knowledge, techniques, attitudes and personal qualities and will include the ability to plan, execute and evaluate digital actions in the solution of life tasks;

v. It also includes the ability to be aware of oneself as a digitally literate person, and to reflect on one's own digital literacy development.

The following definition can therefore be proposed: Digital Literacy is the awareness, attitude and ability of individuals to appropriately use digital tools and facilities to identify, access, manage, integrate, evaluate, analyze and synthesize digital resources, construct new knowledge, create media expressions, and communicate with others, in the context of specific life situations, in order to enable constructive social action; and to reflect upon this process.

Levels of Digital Literacy

Just as we saw above that literacy can be conceived on three levels, we may approach digital literacy in the same vein, seeing it as operative first at the level of technique, of the mastery of digital competences, secondly at the level of thoughtful usage, of the contextually-appropriate application of digital tools, and thirdly, at the level of critical reflection, of the understanding of the transformative human and social impact of digital actions (Figure 7.1). I suggested above that approaches to computer literacy have evolved towards encompassing all three levels. The implication of the definition adopted is that we can only talk about digital *literacy* at levels II or III; digital competence is a requirement for and precursor of digital literacy, but it cannot be described as digital literacy.

Figure 7.1: Levels of Digital Literacy.

Level III: DIGITAL TRANSFORMATION (innovation/creativity)

Level II: DIGITAL USAGE (professional/discipline application)
DIGITAL LITERACY

Level I: DIGITAL COMPETENCE (skills, concepts, approaches, attitudes, etc.)

Digital literacy is conceived as an attribute of the person in a socio-cultural context; it is an element of that person's identity. In considering the pedagogy of e-learning, Mayes and Fowler (2006, p. 27) argue that:

Just as the field of educational technology has matured from a 'delivery of content' model to one that emphasizes the crucial role of dialogue, so the field of digital literacy, we suggest, should shift its emphasis from skill to *identity*. (Italics original) Digital literacy therefore varies between individuals, as their life situations vary—it is a quality of the person, not an externally-defined threshold to be attained. There is no "one size fits all."

The three stages bear some similarity to the three curriculum dimensions for the development of "e-competences" proposed by the EC-funded I-Curriculum Project, the *operational*, the *integrating* and the *transformational* curriculum:

Operational Curriculum is learning to use the tools and technology effectively. Knowing how to word-process, how to edit a picture, enter data and make simple queries of an information system, save and load files and so on.

Integrating Curriculum is where the uses of technology are applied to current curricula and organisation of teaching and learning. This might be using an online library of visual material, using a virtual learning environment to deliver a course or part of a course. . . .

Transformational Curriculum is based on the notion that what we might know and how, and when we come to know it is changed by the existence of the technologies we use and therefore the curriculum and organisation of teaching and learning need to change to reflect this. (I-Curriculum, 2004, p. 7)

Level I. Digital Competence

At the foundation of the system is *digital competence*. This will span a wide range of topics and will encompass also a differentiation of skill levels from basic visual recognition and manual action skills to more critical, evaluative and conceptual approaches and will also include attitudes and awarenesses. Individuals or groups will draw upon digital competence as is appropriate to their life situation, and return to it as often as new challenges presented by the life situation change.

The working group on "key competences" of the European Commission "Education and Training 2010" Programme identifies *digital competence* as one of the eight domains of key competences, defining it as "the confident and critical use of Information Society Technologies for work, leisure and communication." (European Commission, 2004, p. 14) Information society technologies (IST) are defined as "offering services based on the use of Information and Communication technologies (ICT), the internet, digital content, electronic media, etc., via for example a personal computer, a mobile telephone, an electronic banking machine, an *e*Book, digital television, etc." (loc. cit.) Digital competence is regarded as consisting of knowledge, skills and attitudes.

A problem here is the varying meaning of the terms *skill* and *competence*. *Skill* is sometimes seen as representing only lower order attributes (e.g., *keyboard skills*) but sometimes as including also higher order attributes (e.g., *thinking skills* or *analytical skills*). *Competence* (or competency) is sometimes construed as the application of skills in specific contexts but is also seen as synonymous with *skill* or sometimes with higher level skills. The Key Competences working group addresses this issue:

> The terms 'competence' and 'key competence' are preferred to 'basic skills' which was considered too restrictive as it was generally taken to refer to basic literacy and numeracy and to what are known variously as 'survival' or 'life' skills. 'Competence' is considered to refer to a combination of skills, knowledge, aptitudes and attitudes, and to include the disposition to learn in addition to know-how. (ibid., p. 3)

Focusing on generic aspects of transferable "key competences," the working group makes clear that the key competences will enable successful life action:

> Key competences should be **transferable**, and therefore applicable in many situations and contexts, and **multifunctional**, in that they can be used to achieve several objectives, to solve different kinds of problems and to accomplish different kinds of tasks. Key competences are a **prerequisite** for adequate personal performance in life, work and subsequent learning. (ibid., p. 6) (emphasis in original)

We can regard digital competence, as conceptualized in the work of the Key Competences working group, as an underpinning element in digital literacy. In moving from competence to literacy, however, we take on board the crucial importance of *situational embedding*. Digital literacy must involve the successful usage of digital competence within life situations.

We have ordered digital competence around thirteen processes (see Figure 7.2). These are more-or-less sequential functions carried out with digital tools upon digital resources of any type, within the context of a specific task or problem. The problem or task may be in any area of activity: e.g., writing an academic paper, preparing a set of photographs, making a multimedia presentation, or investigating one's family tree. "Digital resources" are to be considered in the most inclusive way, and a digital resource could be defined as any item which can be stored as a computer file. This could include text, images, graphics, video, music, and multimedia objects; digital resources could take the specific form of reports, academic papers, fiction, pieces of music, art works, films, games, learning materials, data collections, etc. The first and last processes, statement and reflection, have a more generic status as mediating processes between digital actions and their cultural context.

Figure 7.2: Processes of Digital Literacy.

Process	Descriptor
Statement	To state clearly the problem to be solved or task to be achieved and the actions likely to be required
Identification	To identify the digital resources required to solve a problem or achieve successful completion of a task
Accession	To locate and obtain the required digital resources
Evaluation	To assess the objectivity, accuracy and reliability of digital resources and their relevance to the problem or task
Interpretation	To understand the meaning conveyed by a digital resource
Organization	To organize and set out digital resources in a way that will enable the solution of the problem or successful achievement of the task
Integration	To bring digital resources together in combinations relevant to the problem or task
Analysis	To examine digital resources using concepts and models which will enable solution of the problem or successful achievement of the task
Synthesis	To recombine digital resources in new ways which will enable solution of the problem or successful achievement of the task
Creation	To create new knowledge objects, units of information, media products or other digital outputs which will contribute to task achievement or problem solution
Communication	To interact with relevant others whilst dealing with the problem or task
Dissemination	To present the solutions or outputs to relevant others
Reflection	To consider the success of the problem-solving or task-achievement process, and to reflect upon one's own development as a digitally literate person

Elements of digital competence for the individual involve instantiations of the processes in a relevant domain. They could therefore include such skills as finding information on the web, word processing and document preparation, electronic communication, creation and manipulation of digital images, use of spreadsheets, creation of presentations, publishing on the web, creation and use of databases, simulations and modeling, desk top publishing, digital and interactive games, production of multimedia objects, and mastery of digital learning environments. Instantiations of digital competence will vary from person to person as their situations vary and will change over time as new tools and facilities are developed.

Components of digital competence may be mastered at levels of expertise which will vary from basic skills to more demanding evaluative or analytical competence. Attempts to define multiple levels of differentiation have not been successful, becoming bogged down in the niceties of defining the exact differences between one level and the next, and it is probably only necessary to have a small number of clear levels.

Level II. Digital Usage

The central and crucial level is that of *digital usage*: the application of digital competence within specific professional or domain contexts, giving rise to a corpus of digital usages specific to an individual, group or organization. In generating digital usages, users draw upon relevant digital competences and elements specific to the profession, domain or other life-context. Each user brings to this exercise his/her own history and personal/professional development.

Digital usages are shaped by the requirements of the situation: they are focused upon solution of a problem, completion of a task, or achievement of some other outcome within the professional, discipline or other domain context. They are thus crucially shaped by the professional, discipline or domain expertise of the individual, without which they cannot be successful. The drawing upon digital competence is determined by the individual's existing digital literacy and the requirements of the problem or task. Digital usages are therefore fully embedded within the activity of the professional, discipline or domain community. They become part of the culture of what Wenger has called "communities of practice":

> Communities of practice are groups of people who share a concern, a set of problems, or a passion about a topic, and who deepen their knowledge and expertise in this area by interacting on an ongoing basis. (Wenger et al., 2002, p. 4)

In communities of practice, learning becomes a communal activity intimately linked with everyday practice. Digital usages become embedded within the understandings and actions which evolve within the community and cause the community itself to evolve; hence, the community of practice is also a community of learning.

Figure 7.3: Digital Literacy in Action.

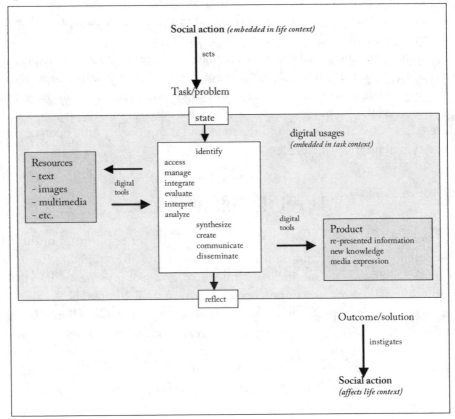

The process in which digital literacy is put into action is shown in Figure 7.3. The task or problem arises out of the individual's life context; it may concern work, study, leisure, or any other aspect of the life context. In order to complete the task or to solve the problem, the individual identifies a competence requirement. He/she may then acquire the needed digital competence through whatever learning process is available and preferred. He/she can then make an appropriate use of the acquired digital competence; this takes place within the context of the task, and is therefore informed and shaped by the knowledge and expertise pertaining to the professional, discipline or other domain context. The informed uses of digital competence within life-situations are termed here *digital usages*. These involve using digital tools to seek, find and process information and then to develop a product or solution addressing the task or problem. This outcome will itself be the trigger for further action in the life context.

Level III. Digital Transformation

The ultimate stage is that of *digital transformation* and is achieved when the digital usages which have been developed enable innovation and creativity and stimulate significant change within the professional or knowledge domain. This change could happen at the individual level or at that of the group or organization. Whilst many digitally literate persons may achieve a transformative level, transformation is not a necessary condition of digital literacy. Activity at the level of appropriate and informed usage would be sufficient to be described as digitally literate.

Users do not necessarily follow a sequential path at each stage. They will draw upon whatever is relevant for the life-project they are currently addressing; the pattern is more one of random rather than serial access, although there will be many cases where certain low level knowledge and skill are necessary in order to develop or understand material from a higher level.

Conclusion

I have hoped to move discussion of digital literacy—and of the literacies that make it up or that relate to it—from the area of listing of skills to be mastered towards that of the role of the digital in the growth of the individual, as student, as worker, as person. Whilst the awareness of skills or competences gained is necessary, it is only a part of the process of achieving study, career and life goals through the appropriate use of digital means. Furthermore, digital literacy is itself an element in the ongoing construction, in a social context, of individual identity.

In the digitally infused world of late modernity, identity is fragile, and subject to many pressures and subterfuges, from those who know, or think they know us personally—friends, colleagues and family—and from those who do not know us personally but wish to know us intimately as constructions of data—business, the media, and the state. We live in a dangerous world, where the collection, collation, mining, exchange and sale of personal data enables the owners of powerful machines to believe that they know us better than we do ourselves, that our uncertainties can be corrected by their certainties. Part of being digitally literate is to be aware of and to resist the digital threats to identity and to be able to use digital means to secure and support one's own identity.

Individuals do not stand alone but are part of society, and individual acts and identities made and remake daily the social order. Social order enables structured social activity to take place which allows individual actions, meanings and identities to coalesce with others and to shape the structures of meaning perceived to be bigger than the individual, to pertain to groups or people, or to society as a whole. Thus, for individuals to view themselves as developing digital literacy and to reflect on the implications of that for their identity and their life plays a part in helping to build socio-cultural patterns which give people some understanding and sense of control in an unstable age.

References

ACRL. (2000). *Information literacy competency standards for higher education.* Chicago, Illinois: Association of College and Research Libraries.

Bauman, Z. (2000). *Liquid modernity.* Cambridge: Polity Press.

———. (2001). *The individualized society.* Cambridge: Polity Press.

———. (2005). Education in liquid modernity. *Review of Education, Pedagogy, and Cultural Studies, 27,* 303–317.

Beck, U. (1992). *The risk society.* London: Sage.

Beck, U., & Beck-Gernsheim, E. (2001). *Individualization.* London: Sage.

Bélisle, C. (2006). Literacy and the digital knowledge revolution. In A. Martin & D. Madigan (Eds.). (2006). *Digital literacies for learning* (pp. 51–67.). London: Facet.

Brzezinski, Z. (1970). *Between two ages: America's role in the Technetronic Era.* New York: Viking Press.

Dakers, J. (Ed.) (2006a). *Defining technological literacy.* New York: Palgrave Macmillan.

———. (2006b). Towards a philosophy *for* technological education. In J. Dakers (Ed.). *Defining technological literacy* (pp. 145–158). New York: Palgrave Macmillan.

Dertouzos, M.L., & Moses, J. (Eds.) (1979). *The Computer Age: A twenty-year view.* Cambridge, Massachusetts: MIT Press.

DfEE. (1998). *The learning age: a renaissance for a new Britain.* London. Stationery Office Cm 3790.

Dondis, D. A. (1973). *A primer of visual literacy.* Cambridge, MA: MIT Press.

ETS. (2002). *Digital transformation: A framework for ICT literacy*. Princeton, NJ: Educational Testing Service.

European Commission. (1994). *Europe and the global information society*. Recommendations to the European Council by the High-Level Group on the Information Society. Brussels.

———. (2003). *eLearning: Better eLearning for Europe*. Brussels: Directorate-General for Education and Culture.

———. (2004). *Key competences for lifelong learning: A European reference framework*. Directorate-General for Education and Culture. Retrieved December 5, 2007, from http://europa.eu.int/comm/education/policies/2010/doc/basicframe.pdf

Feenberg, A. (1999). *Questioning technology*. New York: Routledge.

Flood, J., Lapp, D. & Heath, S. B. (Eds.) (1998). *Handbook of research on teaching literacy through the communicative and visual arts*. International Reading Association, New York: Macmillan.

Forrester, T. (Ed.) (1980). *The microelectronics revolution*. Oxford: Basil Blackwell.

Gates, B. (1995). *The road ahead*. London: Viking Penguin.

Giddens, A. (1990). *Consequences of modernity*. Cambridge: Polity Press.

———. (1999). Risk and responsibility. *Modern Law Review, 62 (1)*, 1–10.

Gilster, P. (1997). *Digital literacy*. New York: John Wiley

Handel, S. (1967). *The electronic revolution*. Harmondsworth: Penguin.

Hobbs, R. (1998). Literacy in the information age. In J. Flood, D. Lapp, & S. Brice Heath (Eds.), *Handbook of research on teaching literacy through the communicative and visual arts* (pp. 7–14). International Reading Association, New York: Macmillan.

House of Lords (1996). *Information Society: Agenda for Action in the UK*. Select Committee on Science and Technology Session 1995–96. 5th Report London. HMSO HL Paper 77.

I Curriculum (2004). Guidelines for emergent eCompetences at schools. I-Curriculum Project Socrates Programme—Minerva Action. Retrieved November 23, 2007, http://promitheas.iacm.forth.gr/i-curriculum/Assets/Docs/Outputs/Guidelines%20Teachers.pdf

ITEA (1996). *Technology for all Americans*. Reston, VA: International Technology Education Association.

Johnston, B. & Webber, S. (2003). Information literacy in higher education: A review and case study. *Studies in Higher Education, 28*, 335–352.

Kahn, R. & Kellner, D. (2006). Reconstructing technoliteracy: A multiple literacies approach. In J. Dakers (Ed.). *Defining technological literacy* (pp. 253–273). New York: Palgrave-Macmillan.

Kellner, D. (2002). Technological revolution, multiple literacies, and the restructuring of education. In I. Snyder (Ed.), *Silicon literacies* (pp. 154–169). London: Routledge.

Kress, G. (2003). *Literacy in the new media age*. London: Routledge.

Lankshear, C., & Knobel, M. (2003). *New literacies: Changing knowledge and classroom learning*. Buckingham: Open University Press.

Large, P. (1980). *The micro revolution*. Glasgow: Fontana.

Martin, A. (2003). Towards e-literacy. In A. Martin & R. Rader (Eds.). *Information and IT literacy: Enabling learning in the 21st century* (pp. 3–23). London: Facet.

Martin, A., & Madigan, D. (Eds) (2006). *Digital literacies for learning*. London: Facet.

Martin, A., & Rader, R. (Eds.). (2003). *Information and IT literacy: Enabling learning in the 21st century*. London: Facet.

Mayes, T., & Fowler, C. (2006). Learners, learning literacy and the pedagogy of e-Learning. In A. Martin & D. Madigan (Eds.). *Digital literacies for learning* (26–33). London: Facet.

Michael, M. (2006). How to understand mundane technology. In J. Dakers (Ed.). *Defining technological literacy* (pp. 49–63). New York: Palgrave Macmillan.

Ministry of Education (2003). *Digital horizons: Learning through ICT.* New Zealand Ministry of Education. Wellington, Revised Edition, December 2003. Retrieved December 1, 2007 from http://www.minedu.govt.nz/web/downloadable/dl6760_v1/digital-horizons-revision-03.pdf

NRC (1999). *Being fluent with information technology.* Committee on Information Technology Literacy, National Research Council. Washington, DC: National Academy Press.

Rader, H. (2003). Information literacy: A global perspective. In A. Martin, A. & R. Rader (Eds.). *Information and IT literacy: Enabling learning in the 21ˢᵗ century* (pp. 24–42). London: Facet Publishing.

SNAB (2001). Consensus of the SchoolNet National Advisory Board on a Foresight of the Role of Information and Communications Technologies in Learning. Retrieved December 1, 2007, from: http://www.schoolnet.ca/snab/e/reports/Foresight.pdf

Snyder, I. (Ed.). (2002). *Silicon literacies.* London: Routledge.

Søby, M. (2003). *Digital competence: From ICT skills to digital "bildung."* Oslo: ITU, University of Oslo.

Street, B. (1984). *Literacy in theory and practice.* Cambridge: Cambridge UP.

Tyner, K. (1998). *Literacy in a digital world.* Mahwah, NJ: Lawrence Erlbaum.

UNESCO. (2003, September 20–23). *Conference report of the information literacy meeting of experts.* Prague, Czech Republic. Retrieved December 1, 2007, from http://www.nclis.gov/libinter/infolitconf&meet/post-infolitconf&meet/FinalReportPrague.pdf

———. (2006). *Education for all global monitoring report 2006.* Paris, France. UNESCO Publishing. Retrieved December 1, 2007 from http://www.unesco.org/education/GMR2006/full/chapt6_eng.pdf

Van Joolingen, W. (2004). *The PISA framework for assessment of ICT literacy.* Powerpoint presentation. Retrieved December 1, 2007, from http://www.ictliteracy.info/rf.pdf/PISA%20framework.ppt

Waks, L. (2006). Rethinking technological literacy for the global network era. In J. Dakers (Ed.). *Defining technological literacy* (pp. 275–295). New York: Palgrav Macmillan.

Wenger, E., McDermott, R., & Snyder, W. M. (2002). *Cultivating communities of practice.* Boston, MA: Harvard Business School Press.

Wilde, J., & Wilde, R. (1991). *Visual literacy.* New York: Watson-Guptill.

CHAPTER EIGHT

Trajectories of Remixing

Digital Literacies, Media Production, and Schooling

OLA ERSTAD

Introduction

Cultural transformations in recent years have been strongly linked to the developments of technology (Castells, 1996; Jenkins, 2006; Buckingham & Willett, 2006). Such transformations can be seen in the ways we as citizens change our everyday practices; as with the introduction of email systems early in the 1990s that changed fundamentally how we communicate, developments of virtual reality and later on simulations, the development of the World Wide Web as a source for information and in later years the developments of online games and Web 2.0.

One of the key challenges in these developments is the issue of digital literacy. This relates to the extent to which citizens have the necessary competence to take advantage of the possibilities given by new technologies in different settings. In a fundamental way it raises discussions about what it means to be able to "read" and "write" as part of our cultural developments today, understood as interpretation of and access to information and how we communicate and express ourselves. My focus in this chapter is on the implications these developments have for the way we think about education and the institutional practice of schooling.

From these more general considerations there is one aspect of digital literacy that I believe is of special importance; that is the issue of media and content production, especially what I term "trajectories of remixing." This is seen in the developments of Web 2.0 and the increasing number of postings on sites like YouTube, MySpace, Flickr or Facebook, during just a couple of years. We can also see similar trends in new television concepts like "current.com," where it is the audience that produces the content, as well as in television shows like "So You Think You Can Dance" or "Idol." This represents a shift in the role of audience and the impact of production practices.

Remixing activities as an essential part of digital literacy represent processes of change in our schools today, from knowledge development being based on predefined content in school books and the reproduction of knowledge provided by the teacher, towards a situation where students take available content and create something new, something not predefined. Some schools have already implemented these new possibilities provided by digital media, and several theoretical developments are highlighting the educational implications of these developments (Scardamalia & Bereiter, 2006). However, our education system is still fixed in the traditional ideals of literacy.

I will use my own country, Norway, as an example of a country trying to address these challenges through its education system and policy developments. I will start with some contextual information about developments in Norway. The next step will be to outline some issues and frameworks about digital literacy in the research literature. From this I will orient myself more towards discussing the concept of remixing, and advancing my point that this has to be seen over longer processes of media production and not only as an expression of putting different content pieces together. I will use three cases as examples of such processes within school-based settings. These are taken from different projects in Norway where remixing activities are part of project work in schools using digital media. Towards the end I will look at some of the implications this has for our conceptions of schooling.

The Norwegian Context

Developments in Norway during the last 10 years can be divided into three main phases indicating the overall national agenda for scaling up activities using digital media in Norwegian schools. The three phases are also expressed in specific "action plans" from the Ministry of Education. The first phase, from 1996 until 1999, was mainly concerned with the implementation of computers into Norwegian schools. There was less interest in the educational context. In the next phase, from 2000 until 2003, the focus was more on whole school development with ICT and changing learning environments. The phase we are in now, from 2004 until 2008, puts more emphasis on digital literacy and knowledge building among students and what learners do with technology.

One immediate challenge in these developments has been the balance between "top-down" and "bottom-up" strategies. At one level it has been important to commit the Ministry of Education to developing ICT in Norwegian schools. At another level it has been important to get schools to use ICT more actively. The latter has been more difficult. In the last 3 to 4 years this has changed in the sense that more schools start activities themselves, since access to computers and the internet is no longer a problem either in school or at home.

The new national curriculum from 2006 defines digital literacy/competence[1] and the "skill to use digital tools" as being as important as reading, writing, numeracy and oral skills. The implication is that all students on all levels and in all subjects should use and relate to digital media in their learning processes in Norwegian schools. The emphasis is mainly on skills in using the technology although broader issues of competence such as evaluating sources critically when using the internet and using ICT to collaborate are also implied.

At the same time research has shown that even though the access to digital media and the internet has been steadily improving, teachers hardly use these media in their activities with students. The students report that they use such media much more at home, and for a broader scope of different purposes (Erstad et al., 2005).

The conception of "digital competence" has been important on a policy level to create more awareness of the impact of digital technologies on our education system. In a Norwegian setting we are now in a situation where the main question is what students and teachers use these new technologies for, both inside and outside of schools. During the last few years many initiatives have been taken to stimulate productive use of new technologies in schools, as in numerous other countries. This has often been informed by how young people use new technologies outside of schools.

Norwegian youth, like youth in many other countries, are very active users of new digital media. Television is still the medium most youth use on an average day, but the time spent on the traditional mass media like television, radio and newspapers (paper versions) has been steadily declining since 2000. Most young people use a broad range of different digital media. Compared to young people in some other European countries, proportionately more Norwegian youth fall into the category of "advanced users" (Heim & Brandtzæg, 2007). This implies that Norwegian youth have good access to technology and use such technologies for "advanced" purposes. However, in a general sense several studies have shown how young people gain most of their competence in using digital technologies outside the formal institutions of knowledge building (Livingstone, 2002; Buckingham, 2003; Alvermann, 2002).

It should also be mentioned that young people in the Nordic countries have been involved in online networking sites and online content production for several years. In all Nordic countries there are examples of online communities for social networking and media production by youth since the end of the 1990s, which have been very popular. Most famous is the Swedish "Lunar Storm."

What is interesting is that these developments in policy initiatives towards digital literacy and media use in general have brought Norwegian educational initiatives to a point where new cultural practices start to appear, making us look at literacy in new ways. Remixing is such a turn in the way digital media allow the user to create new content based on other peoples' content production and sharing it with others. What is important is to frame remixing as such within broader discussions about digital literacy.

Digital Literacies: Conceptions, Frameworks and Issues

In recent years there has been an interest in how traditional conceptions of literacy change due to new digital technologies. An important point is that literacies change over time due to socio-cultural processes (Scribner & Cole, 1981; Olson & Cole, 2006). Similar perspectives are reflected in socio-cultural theories of learning, where learning is related to the use of specific artifacts and tools (Säljö, 1999). James Wertsch (1998) uses concepts like "cultural tools" and "mediational means" to discuss these transitions in human development:

> One could focus on the emergence and influence of a new mediational means in sociocultural history where forces of industrialization and technological development

come into play. An important instance of the latter sort is what has happened to social and psychological processes with the appearance of modern computers. Regardless of the particular case or the genetic domain involved, the general point is that the introduction of a new mediational means creates a kind of imbalance in the systemic organization of mediated action, an imbalance that sets off changes in other elements such as the agent and changes in mediated action in general. (Wertsch, 1998, p. 43)

Similar ideas are expressed by the German literary scholar and media theorist Friedrich Kittler who described this as the development of different "cultural techniques" over time (1990). Wertsch (1998) also makes a point of distinguishing between mastery and appropriation in relation to different cultural tools. Concerning digital literacy, this relates to being able to operate the technology itself, to master it, versus having the competence to reflect on the use of digital media in different contexts as part of your identity as a learner.

This implies that we constantly have to keep in mind the more general question of what it means to "read" and "write" in a culture and, thereby, how we learn (Pahl & Rowsell, 2005). In the *Handbook of Literacy and Technology*—subtitled, "Transformations in a Post Typographic World"—David Reinking and colleagues (1998) present several perspectives on how the development of digital technologies changes conceptions of text, of readers and writers and ultimately of literacy itself. This implies that digital literacy relates to changes in traditional cultural techniques like reading and writing, and yet meanwhile opening up new dimensions to what it means to be a competent reader and writer in our culture.

In her book, *Literacy for Sustainable Development in the Age of Information* (1999), Naz Rassool presents an overview of different debates on literacy during recent decades. Her point is that research perspectives on technology and literacy need to reconceptualize power structures within the information society, with an emphasis on "communicative competence" in relation to democratic citizenship. Digital technologies create new possibilities for how people relate to each other, how knowledge is defined in negotiation between actors and how it changes our conception of learning environments in which actors make meaning. Empowerment is related to the active use of different tools, which must be based upon the prerequisite that actors have the competence and critical perspective on how to use them for learning. Literacy, seen in this way, implies processes of inclusion and exclusion. Some have the skills and know how to use them for personal development, others do not. Schooling is meant to counteract such cultural processes of exclusion.

What exactly should be included within the conceptual domain of literacy has become increasingly fuzzy, especially among those educators and researchers whose professional interests are tied to how literacy is understood. This is, of course, due to the fact that literacy is not a static term but relates to technological innovations, and cultural and political strategies and developments. It is necessary to distinguish between more of a skills orientation and a higher level of competency.

Table 8.1. Key Concepts of ICT Literacy (my elaboration based on key concepts in the ETS *Digital Transformations* report, 2002)

Basic skills	Be able to open software, sort out and save information on the computer, and other simple skills in using the computer and software
Download	Be able to download different information types from the internet
Search	Know about and how to get access to information
Navigate	Be able to orient oneself in digital networks, learning strategies in using the internet
Classify	Be able to organize information according to a certain classification scheme or genre
Integrate	Be able to compare and put together different types of information related to multimodal texts
Evaluate	Be able to judge the quality, relevance, objectivity and usefulness of the information accessed. Critical evaluation of sources
Communicate	Be able to communicate information and express oneself through different mediational means
Cooperate	Be able to take part in net-based interactions and learning and take advantage of digital technology to cooperate and be part of different networks
Create	Be able to produce and create different forms of content: generate multimodal texts, make web pages, and so forth. Be able to develop something new by using available tools and software

One report often referred to is *Digital Transformations. A framework for ICT Literacy* (ETS, 2002) written by a team of experts for the Educational Testing Service in the U.S. In this report they identified some key concepts of what they called ICT literacy. One interpretation of such key concepts is presented in Table 8.1 (which comprises my own elaboration of key concepts in this ETS report).

This consists of more general competencies that are not connected to specific subjects in school or specific technologies. They can be taught and are related not only to what is learned in school settings but also to situations outside the school.

Other frameworks have used "digital competence" as an overall term. One example is the working group on "key competences" of the European Commission's "Education and Training 2010." This program identifies *digital competence* as one of the eight domains of key competences; that is, as "the confident and critical use of Information Society Technologies for work, leisure and communication. These competences are related to logical and critical thinking to high-level information management skills and to well-developed communication skills. At the most basic level, ICT skills comprise the use of multi-media technology to retrieve, assess, store, produce, present and exchange information, and to communicate and participate in networks via the internet" (European Commission, 2004, p. 14). Digital competence in this framework encompasses knowledge, skills and attitudes related to such technologies.

Another interesting conceptual background is the term media literacy (Buckingham, 2003), which has been part of the media education movement from the 1980s. Discussions on media literacy during the last ten years, especially in the UK (ibid.), are relevant here; first, because the media themselves are the object of analysis; second, because the reflective and critical dimensions of analysis are central; and third, because media production among the students is a key component (see Case 3, p. 195).

Kathleen Tyner (1998), who uses media education and media literacy as a reference in her discussions, studies some of the elements of a modern interpretation of literacy both related to what she terms "tool literacies," to indicate the necessary skills to be able to use the technology, and "literacies of representation," to describe the knowledge of how to take advantage of the possibilities that different forms of representation give the users. This is to state a division between a tool orientation of literacy and a more reflective social process.

An important cultural development in recent years has been the processes of convergence (Jenkins, 2006). This relates to how technologies merge, how production of content changes, how new text formats are developed, and how the users relate to information as part of communication networks in different ways. Parallel to such convergence processes some literacy theorists have sought to hold together the many new literacies under some umbrella concepts stressing the plurality of literacies, such as "multiliteracies" (Cope & Kalantzis, 2000; Snyder, 2002) and "metamedia literacy" (Lemke, 1998). According to Kellner (2002, p. 163), "The term 'multiple literacies' points to the many different kinds of literacies needed to access, interpret, criticise, and participate in the emergent new forms of culture and society." Kress (2003) however argues against the multiplicity of literacies, suggesting that it leads to serious conceptual confusion. He believes that instead of taking this path, it is necessary to develop a new theoretical framework for literacy which can use a single set of concepts to address the various aspects of literacy.

In addition, it is important to stress that technology literacy is related to *situational embedding*, that is, the use of technology within life situations. To understand such processes we have to look at different contexts where literacy is practiced and given meaning. This is especially important when relating it to how children and young people use digital technologies across contexts. In line with this perspective, Lankshear and Knobel (2006) have defined literacy in this sense as:

> Socially recognized ways of generating, communicating and negotiating meaningful content through the medium of encoded texts within contexts of participation in Discourses (or, as members of Discourses). (Lankshear & Knobel, 2006, p. 64)

This definition is not bound by certain technologies. It proposes to study literacies in practice (what people do with technologies and digital texts) and not as something pre-described, indicating that we need to understand what people are already practicing concerning technological literacies and what the role of education should be in employing such literacies for new knowledge levels. The important message is that digital competence among young people today is of direct relevance to discussions about learning in schools, and it seriously confronts earlier conceptions of literacy and learning.

As shown in this section, there are different frameworks to relate to in our understanding of digital literacy/competence. However, the key challenge is to go deeper into the implications of increased use of new technologies in educational practices. I believe the concept of remixing points us in the right direction, because it raises some key issues of educational work with digital media.

Conceptual Developments of Remixing

As several authors have pointed out (Lankshear & Knobel, 2006; Manovich, 2007), the conceptual understanding of remixing is nothing new. Issues of re-using and reworking from other texts have been known since the early days of the Greeks. In more recent times we see such issues expressed in the developments of more visual media, in painting using inspiration from other art, or everyday objects put into paintings, creating new contexts for interpretation—for example, as seen in the Dada movement—or in photography as seen in collage and photomontage (Ades, 1986) or in film as seen in the theories and films of Eisenstein on montage, putting images and scenes together in specific ways, creating new ways of interpretation (Bordwell, 2005).

The concept of remixing is first and foremost connected to developments of producing music through available mixing equipment and in the way DJs work. As stated by Manovich (2007), "[r]emixing originally had a precise and a narrow meaning that gradually became diffused. Although precedents of re-mixing can be found earlier, it was the introduction of multi-track mixers that made remixing a standard practice." (no page) Since the mid-1970s we have seen many examples of how artists take existing music pieces or recordings and make something new from them. "Gradually the term became more and more broad, today referring to any reworking of already existing cultural work(s)." (Manovich, 2007, no page)

For my purpose here it is especially developments of new digital technologies and the way these technologies have become available both at home and in schools that are interesting. As a consequence, remixing as cultural practice has changed dramatically in recent years. Digital tools create new possibilities for getting access to information, for producing, sharing and reusing. The main point is that more and more people in our culture can take part in these re-mixing activities; not only an elite or specific groups. Most evident, it is young people who take a lead in creative practices using digital media.

Some even talk about a remixing *culture* (Lessig, 2005; Manovich, 2007) as characteristic of the changes we see in our culture today. Remix is seen as evident in every domain of cultural practice (Lankshear & Knobel 2006, p. 177). Everyone engages in remix in this general sense of the idea, and remix is everywhere, and defined as a condition for cultural development. What is new is, of course, the impact of digital technologies. The possibilities of remixing all kinds of textual expressions and artifacts have thereby changed. And as mentioned before, these kinds of practices have become central to the ways young people make meaning and express ideas. In his writings on copyright issues and intellectual property legislation in the digital age (as seen in the guidelines on Creative Commons) Lawrence Lessig (2005; see also Chapter 12 here) highlights remix as a key rallying point. For Lessig digital remix constitutes a contemporary form of writing.

Several writers have been looking at remixing as new ways of conceiving text production. Manovich (2007, no page) sees developments of remixing in "music, fashion, design, art, web applications, user created media, food"—cultural arenas that "are governed by remixes, fusions, collages, or mash-ups." Moreover, "if post-modernism defined 1980s, remix definitely dominates 2000s, and it will probably continue to rule the next decade as well."

One aspect of the development of remixing as a concept and practice is the theoretical development of multimodality, especially by scholars like Gunther Kress, Theo van Leeuven and Carey Jewitt. Multimodality expresses the combination of different media elements into a new textual expression. This combination of media elements is not just a sum of the different elements but creates something new, a new quality as text. Some also talk about this as "remediation" (Bolter & Grusin, 1999), as partly building on what exists to develop something new, and partly that digital texts represent something new as hypertext (Landow, 2006). However, this literature has paid little attention to the dynamic process of media production made possible by digital tools.

The Bricoleur of Remixing

As stated in my Introduction, my main argument in this chapter is that "digital literacies" *per se* does not tell us very much about these new cultural practices mentioned above. How digital technologies, for example, influence educational practices needs to be specified through the activities students and teachers are involved in, and this is where remixing becomes interesting and an important facet of digital literacy. This implies a conceptual understanding of remixing that involves the actor to a larger degree than stated in the section above.

Together with some colleagues I have defined re-mixing as "selecting, cutting, pasting and combining semiotic resources into new digital and multimodal texts (bricolage), which is achieved by downloading and uploading files from different sources (internet, iPod, DV-camera, digital camera or sound recording devices)" (Erstad, Gilje, & de Lange, 2007, no page). This implies a focus on the process of remixing and text production.

Studying such processes can be traced back to the concept of "bricolage" from the cultural studies tradition, mainly associated with the Birmingham Centre for Contemporary Cultural Studies during the 1970s. Dick Hebdige, in his classic book, *Subculture: The Meaning of Style* (1979/1985), used the concept of "bricolage" to discuss cultural practices that different youth sub-cultures were involved in at the end of the 1970s, especially the expression of style. This constitutes processes of signmaking done by people in specific sub-cultural settings, called bricoleurs. One of the groups Hebdige was studying was the Mods, characterized by their scooters, grubby parka anoraks, and so forth. According to Hebdige, " . . . the mods could be said to be functioning as bricoleurs when they appropriated another range of commodities by placing them in a symbolic ensemble which served to erase or subvert their original straight meanings" (1979/1985, p. 104). Similar sign-using practices can be seen in how the punk movement used swastikas, rubbish, and safety pins in the face, excessive hairstyles, and so forth to create new interpretations of traditional meaning making from specific signs.

This also raises the question of authoring in remixing activities. In school-based activities the question of copy and paste has been raised as a concern since students have been said to just take elements from other texts and copy them into their own texts without much reflection. However, research that has been done on these activities shows that if we look at this in longer trajectories of activities we find both discussions and reflections on the selection, implementation and expression of different media elements into new textual expressions by students (Rasmussen, 2005). Multimodal practices could be said to give young people a voice to express their positions and interests as agents of remixing. This can be seen in several initiatives about digital storytelling and self-representation using digital tools, where these activities with young people often are defined outside of school-based settings in order to avoid the contextual constraints of schools and build directly on everyday experiences with technologies (Hull & Greeno, 2006).

Lessig (2005) refers to a particular practice of creative writing within the school curriculum in parts of the U.S., where students read texts by multiple authors, take bits from each of them, and put them together in a single text. This is described as "a way of creating something new" (ibid.). Lankshear and Knobel (2006) relate this perspective from Lessig more specifically to issues of literacy in the sense that learning to write is done "by doing it."

For most adults the act of writing means writing with letters, while for young people writing today means something much more using images, sound and video to express ideas (ibid., p. 177–178). In their discussions of remixing, Lankshear and Knobel (ibid., p. 178) include both "practices of producing, exchanging and negotiating digitally remixed texts, which may employ a single medium or may be multimedia remixes," and "various practices that do not necessarily involve digitally remixing sound, image and animations, such as fanfiction writing and producing manga comics."

How can we then understand these trajectories or acts of remixing, and how do they relate to literacy and learning? Nicholas Diakopoulos (2005) has developed an illustration of different acts of remixing combining media elements/pieces and the person involved (see Figure 8.1).

Figure 8.1. Graph Representation of Different Modes of Remix as They Relate to People and Media Elements. (Diakopoulos, 2005, no page)

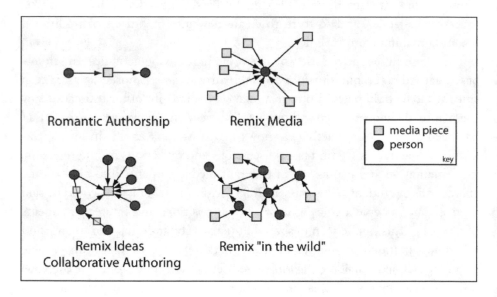

The "romantic authorship" is what we traditionally think about as the writer creating the text: that is, "the author as a lone creative genius." The person/author writes a text, the media piece, which is read by another person. Diakopoulos' point is that this is more of a romantic conception of the writer than what the real situation of authorship has been. This other conception he describes as "collaborative authorship," which has been central throughout history. Diakopoulos suggests we think of "the myriad of different traditional productions which rely on the creative input of multiple people: orchestra, film production and] architecture" (Diakopoulos, 2005, no page). He goes on to describe how "[t]his notion is reflected in Barthes' argument that a text does not release a single meaning, the 'message' of the author, but that a text is rather a 'tissue of citations' born of a multitude of sources in culture (Barthes, 1978). In this light, the author is simply a collaborator with other writers, citing them and reworking their ideas" (Diakopoulos, 2005, no page). In this sense, Diakopoulos distinguishes between remix ideas and remix media. "Remix media" implies that the remixer starts with concrete instantiations of media that are then segmented and recombined, as putting different elements together. Further, "remix ideas" may involve one or more people combining ideas gleaned from different sources (i.e., interpretations of media), which are then potentially instantiated in media, as by bringing ideas together for developing a specific text. "Remix in the wild" can then be related to Web 2.0 and the way media production is done through different people creating different media pieces, which are then reworked by other people through new steps of media production in a complex remixing trajectory.

Mimi Ito has written extensively about such complex cultural production practices, specifically on how Japanese youth are involved in "media mixes" in different ways. Her perspective is that "digital media broaden the base of participation in certain long standing forms of media engagement. This includes the growing accessibility to tools of media production, as well as more diverse internet-enabled means for communicating about and trafficking in cultural content" (Ito, 2006, p. 50). She uses Japanese animation media mixes as an example of combinations of various analog and digital media forms. She argues that "children's engagement with these media mixes provides evidence that they are capable not only of critical engagement and creative production, but also of entrepreneurial participation in the exchange systems and economies that they have developed around media mix content" (ibid.). She then goes on to show how these practices represent a media literacy that young people are involved in as part of their everyday culture.

Another interesting example of such creative media production among youth can be seen in the not so publicly known free online software called "Scratch" (http://scratch.mit.edu/) (Peppler & Kafai, 2007). It has been made to stimulate kids in creative processes, to use a set of available modules, like Legos, to build a short animation, or create something new using available programming possibilities, download animations that others have done, and/or upload your own to the website. When I logged on today (October 12th, 2007) there were available 34,991 projects with a total of 537,575 scripts and 192,959 sprites created by 7,506 contributors among 40,036 registered members. One of the uploaded animations I found on the front page was called "Stick Fight Remix"by Abudeok. This was a simple animation showing two characters fighting and flying between buildings, clearly referencing to computer games and tv cartoons (e.g., *The Powerpuff Girls*). Towards the end text-boxes are included indicating a communication between the two characters, and then ending the short film with the message "The End. Want a part 2? Put it in my comments :)".

Remixing, in this sense, illustrates how young people today are involved in taking different media extracts and putting them together in new ways. My issue here is to show how these new possibilities of media production, expression and cultural practice of reading and writing change how we perceive literacy and learning within school-based settings and how digital literacy will be understood and worked on within these settings.

Remixing Activities in Schools

Media production has a long history in schools, through traditional writing activities, and later on (1980s and 1990s) with audio-visual media in media studies and during more recent years with the massive introduction of computers and internet access (Buckingham, 2003, 2007). The question that arises is how media production practices in schools today indicate a change of what it means to read and write, building on the experiences of young people from outside of schools, and the importance of media production through all subjects and levels and not only as part of specific media-related subject domains.

Schooling in this sense refers to the process of relating to all aspects of being at schools as we usually know them today. Several studies have shown that there is a gap between how much and for which purposes digital media are used in- and outside of schools (Erstad et al., 2005). At school, children report that they use digital media to a much lesser degree than at home or with friends, and that when they use such media at school it is often to make traditional literacies of writing letters and numeracy more effective, while their use outside of schools is more made up of many different activities, especially gaming, downloading music and communication and creation. There are, however, some schools that have managed to create new spaces for literacy and learning, taking advantage of the possibilities represented by digital media, often connected to project work (Erstad, 2005).

Three Cases

Below I will briefly describe three cases where I see that trajectory of remixing, as part of digital literacy, being expressed. These are taken from different projects I have been involved in and are not unique. Many similar examples could have been mentioned. An interesting development in Norwegian schools is, for example, the introduction of digital storytelling as a method of telling personal stories remixing images, music, sound, visual effects and voice over (for more on this, see Erstad & Silseth, 2008). The first case is taken from one of the Norwegian cases in a large international project focusing on "innovative pedagogical practices using information and communication technologies" (Kozma, 2003, no page). The second is from a small project involving two classes of students at two different schools in Oslo, who collaborated on a project on prejudices and created online newspapers. The third is taken from ongoing research on media studies in Norwegian schools. None of these was initiated to study trajectories of remixing explicitly. Nonetheless, they raise issues of how this now becomes part of digital literacies in schools, showing both possibilities and constraints. I will not analyze the interaction among the students in these projects in an empirical sense. The intention here is to describe and reflect on the activities themselves related to the way they illustrate remixing using digital media in schools.

Case 1: Crossing Borders and Modalities

At one lower secondary school just outside Oslo a pair of teachers initiated a project called the "Antarctica project." It all started in October 1999 when the explorers Liv Arnesen (Norwegian) and Ann Bancroft (American) presented their ideas for an education program connected to their Antarctica 2000–2001 expedition. This was presented as a global activity where schools in different countries could participate. A special database was developed where anyone could follow the expedition. In addition the school had a special arrangement with one of the explorers, Liv Arnesen, whereby they would have direct interaction before, during and after the expedition. This was both to get factual and research-based information, and information of a more personal nature about the experiences of the two women in Antarctica. The student project was about the Antarctic.

On using new technologies and related to remixing in the school the principal says that:

It relates to being able to use many senses, and to do things and to see that it works. To learn about another country by reading about it in a book compared to having it presented through internet, images, sound and experiences, you might say, and communication with students in other countries directly through email and chatting and all that which now is possible. (Interview with school principal)

Two teachers and eight students took part in this project. The aim of the project was to create a web page that would contain different kinds of reports and information gathered by the students about the expedition and Antarctica.

The online version of one of Norway's major newspapers was a collaborator in the project. It had a special agreement with the expedition organizers to get up-to-date information. The newspaper also established a link to the students' web page. In addition, the students used the internet to get access to more general information about Antarctica, and they downloaded some video-presentation program and also digital programs to edit the audio interviews with the explorers and put it on their web page.

The main technology used in this project was connected to the creation of a project web site. Additional activities consisted of collecting information from different sources and presenting it on the web site. The web site was created as part of the national school net and thereby became available to all schools in Norway. They had about 3,000 visitors per week. For working on their web page the students used Page Pro, Photoshop and FrontPage 2000. Mainly one PC was used for updating the web site. They used the internet to get access to information and email to stay in contact with the explorers and other students in and outside Norway. They used Word and learned basics of html editing and coding. A couple of the students were reasonably familiar with programming, knowing more than the teacher.

Different kinds of technologies have been used in different phases of the project. It started out with ordinary information retrieval on the internet about Antarctica followed by use of extensive use of email to exchange information with students in other countries. The next step was to create web pages about the expedition. On their web site the students made a digital map on which they plotted the route that the explorers took from week to week. One of the teachers mentioned that he also used SMS messages on the mobile phones to get in touch with the students after school hours. He sent out SMS messages to the students when the satellite connection with the explorers was established, and then all the students came to the school to participate. In addition, as a consequence of the project they started to use video conferencing in their collaboration with other schools.

This project illustrates, in a simple way, processes of remixing in the way students searched for, brought together and combined different media elements made possible by digital technologies. It was also evident that for the students this project created some new perspectives on the school as a knowledge institution. By commenting on the use of technology in such a project some students mention that:

Boy: It becomes more fun to be at school. When you split it up a bit more. Instead of having six hours in one stretch, then it becomes easier to get through the day.

Girl: For some it might be a big shock when they get into the work market, because you do not sit and make mathematical assignments as such. When we work on projects you get a better grasp on what is happening in real companies and such.

Boy: We should get more experience on how it is in real working life.

In relation to this project the technology has given the students some opportunities and arenas for negotiation that creates exciting consequences for how they work on subject matter. As shown, remixing activities are a central part of the whole project integrating different modalities and knowledge domains in the making of the website and following the expedition.

Case 2: Challenging Prejudices of the Other

This case story is taken from a project involving two lower secondary schools, one in the eastern part of Oslo and the other in the Western suburbs. Both schools had long experiences of project work as the main school activity all year round. At each school a group of students took part in the project during a two-week period (approximately 20 students in one school and 40 students in the other). The teachers at the two schools had for some time talked about collaborating on a project focusing on the differences between the two schools. The school in the Western suburbs had students from families with a high socio-economic background with only one non-white student, an adopted child born in Chile. At the school in the Eastern inner city part of Oslo the students came from many different cultural backgrounds with about 65% of the students from minority language-speaking families. After discussing this with the students the teachers decided that the students should collaborate on a project about prejudices concerning east and west in Oslo, and that they should use technology as a central part of the project work. When the project was starting up there were several headlines in the national newspapers about a study showing huge differences in the expected life course and death rate of people living respectively in the east and west of Oslo. This shocked the students and was an important stimulus for their discussions.

In the project the students used different digital tools to collaborate and create an online newspaper, one for each school, which consisted of reportages about the students on the other side of town, their community and their school. In each student group they divided themselves into an editorial board with responsibilities for different sections of the paper; on culture, religion and ethics, sport, statistics about their communities and interviews with inhabitants. They created questions that they sent to each other using a collaborative online platform and MSN. Halfway through the project a group of students from each school traveled, without the teachers, to visit the students at the other school using public transportation. None of the students had ever been in the area of the other school. To document this visit each group made a video film and took still pictures to use in their own production.

My interest here is not to discuss everything that happened in the course of this project or the outcome concerning the prejudices about the Others, which turned out to be a very stimulating process in itself. I use this case as an example of a project where the dynamic use of digital media is integral to the ways students work and to what they create.

Throughout the project the students worked with different modalities and information sources in the making of the online newspapers. They worked individually on different computers looking for images, statistical data, graphs, illustrations, written texts, or editing audio interviews with players from the local soccer team, editing the videofilms to put on the web, and then got together to negotiate how to integrate and remix the different content sources into something new on their online newspaper. The two online papers turned out very differently. One had many different visual effects, with numerous images on the front page and links to other sections of the paper consisting of more text and images. The online paper of the other school had simpler aesthetics on the front page and more video material, such as video interviews with students at their own school and with students from the other school recorded during their visit.

What was interesting in this project were the ways the students combined different content they found on the internet with their own content, either written texts through collaborative writing or audio- and video-tapes. The editorial group at each school had the last word concerning how things should be presented on their online newspaper. Reviewing my video-observations of the two groups revealed a very intense and creative process among the students working on different content materials and sending between the two schools. Often there were rapid changes in the way they related to content materials, such as when one of the Muslim girls at the school in eastern part of Oslo described why she was wearing a veil. This subsequently generated a lot of questions to her from the students at the other school on what this meant in everyday activities like what she did during gymnastics lessons, did she have ethnic Norwegian friends, what were her interests in music or films, and what did her parents think about her growing up in Norway. In documenting this story the students remixed different content materials they found on the internet about the Muslim religion, about world incidents connected to religious conflicts and then connected to this girl's personal story, which was then presented on the online newspapers of both schools. In a simple way this project generated a lot of online and offline discussions about different themes triggered by their own prejudices towards each other and documented through collaborative efforts of remixing different content materials on an online newspaper for each school.

Case 3: Media Production in Media Education

The third case is taken from a subject domain in schools where digital media are at the core, both as embedded in learning activities and as an object of analysis. In the year 2000 a new subject was introduced at the upper secondary level in Norway, called "Media and Communication," as an optional three-year program in vocational training but also with academic components. Media education has been part of Norwegian education for many years but only as a marginal issue comprising a non-compulsory part of school programs. However, this new program has become very popular among students, and more and more schools are establishing it and investing in infrastructure and teacher competencies for offering it.

For my purpose here the interesting part is the strong emphasis on media production in this subject made possible in new ways by digital tools. My interest is not in what they make, the end product but, rather, in the process of media production, what I have termed "trajectory of remixing," using digital tools.

The first example of such remixing activities was collected by one of my Ph.D. students, Øystein Gilje, as part of his research on this subject in Norwegian schools (see also Erstad, Gilje, & de Lange, 2007). The students in this example were working on a short documentary film about a house occupied by young people in Oslo called "Blitz." They were using different digital resources to include in the film employing Photoshop. In the sequence below they are working on the opening title to find the right font, which is very important for them. They did not like the available fonts already in Photoshop, so they went on the internet and found a specific site with alternative fonts where they discovered what they were looking for and then tried to download it to their own computer without succeeding. This is when they asked the teacher to help them.

Teacher: You are going to import a *new* font into Photoshop? (. .).

Girl 1: Yeah. We have downloaded a new font from the internet. That's the problem we want you to help us to solve.

. . . (approx 4 seconds of silence)

Teacher: Well, I don't know how to do that. Why can't you just use one . . . There're plenty of fonts in the program! (scrolling the font-menu in Photoshop)

Teacher: Waste no time doing this! Use time on. (. . .) What's important here is telling the story! You have to work with journalism!

Girl 2: But maybe the font could tell something important about Blitz.

Girl 1: It gives the expression . . . (interrupting each other)

Girl 2: Not use the ordinary one.

The creative element in choosing a particular font for the title is not accepted or is at least hard to understand for the teacher. The confrontation develops further when the teacher tells them that the task is to work with journalism, not with design and "details" in the process. This leads to a discussion about the importance of the downloaded fonts. The students argue that these fonts are important because they express their understanding of the "blitz" concept.

The second example was collected by another doctoral student, Thomas de Lange, as part of his research also on "Media and Communication" (see also Erstad, Gilje, & de Lange, 2007). In this example a student (Boy 1) wishes to supplement his Flash-production with a specific jingle from a Play Station game called *Final Fantasy*. He wants to include this jingle in his production as a personal attribute. Together with a fellow student (Boy 2) he first tries to search for this jingle on the internet. The following extract initiates the overall sequence:

> **Boy 2:** (. . .) who did the music for *Final Fantasy?*
> **Boy 1:** Nobuo Uematsu.
> **Boy 2:** How is it spelled?
> **Boy 1:** N [O U B] ((spelling the name))
> **Boy 2:** [No], say it again.
> **Boy 1:** Let's see. <N O B O U E:::>, <no::u:: bou:: Nubou:: . . . >

The excerpt below takes place about 1 minute later, after the students have found the jingle on the internet and downloaded it to their computer. It starts with Boy 2 playing the jingle loudly on the computer, getting the attention of the teacher who is standing nearby.

> **Teacher:** Quit playing.
> **Boy 2:** . . . was only looking for some music from this game here. I'm not going to play, just getting the music.
> **Boy 1:** I'm not just sitting here watching him play, right?
> **Boy 2:** No! ((Ironic))
> **Boy 2:** A film with Japanese subtitles.
> **Teacher:** That you are going to put into ehh . . . use the music in flash?
> **Boy 2:** Yeah.
> **Teacher:** Uhm.
> ((Teacher walks away))

Both these classroom situations show specific remixing practices among the students as part of larger media production projects. They go on the internet to find a specific font or a jingle that they have clear preferences for and download these to use in their own media productions. These extracts also show how the students operate by themselves and how the teacher has problems dealing with these remixing activities among the students.

Remixing as Literacy

Building on the different conceptions and frameworks of digital literacy discussed earlier in this chapter, especially conceptions of multiliteracies and multimodality, I defined remixing as a key issue in the way digital literacy is developed in contemporary cultural practices. As shown, remixing is nothing new in a cultural sense. What is new, however, is the way digital technologies make it possible to combine many different resources by an increasing number of people. Something new is created based on existing content and then shared with others for further reworking. In this way people take an active part in content production and sharing of multimodal texts. As such it makes a fundamental change in the way we conceive reading and writing as cultural practices of meaning making.

In the cases mentioned above an important aspect of remixing has been the process of creation, what I have termed trajectories of remixing. I have used the context of school to show how this can be played out in a specific setting traditionally framed within the use of books. The cases mentioned are not spectacular in their technology use, and they are representative of what is happening in many classrooms at the moment. What is important is partly that project work is a working method used in many Norwegian schools, which allows students and teachers to work in interdisciplinary ways with a particular theme over time, and partly that there is a good access to technology in Norwegian schools, which makes it more interesting to ask questions about literacy practices involving new technologies. All the cases mentioned above show how available digital tools support the students' remixing activities of using different information sources, combining them in different ways, creating something new, and then sharing this with others for possible reuse. It is the trajectories of remixing that are important in these specific school contexts. At the same time they indicate constraints of doing this in schools. This has not been so obvious in the case descriptions, but through interviewing several of the students involved in these projects they identify clear differences between doing such activities at home and doing them at school. This is mainly connected to the role of the teacher that restricts the students more than supporting or challenging them, partly due to lack of digital competence. This raises questions about what kinds of teacher and student roles such remixing activities open up.

The important challenge is to move away from a simplistic understanding of digital literacy as the skills in operating the technology, towards the more complex set of competencies involved in multiliteracies. Remixing encompasses many of these competencies, such as selecting, organizing, reflecting, evaluating, creating and communicating. And as a literacy it is closely related to the developments of new digital media (Olson & Cole, 2006).

The Idea of Schooling

In an increasing number of documented projects we now see how students use their experiences with creating media content in schools as part of remixing activities. Going from these micro-levels of analysis we can move to more macro levels and see how these developments also challenge the traditional conception of schooling.

My point is that remixing as a cultural activity, especially present in the way young people today use digital media, opens up fundamental questions about "reading" and "writing" and about what schools are for. This creates new conceptions about texts that we read and write, about the student as a producer of content and knowledge, about the roles of teachers and students as part of knowledge-building processes, about identity and learning, and about reproduction versus creation. In the examples mentioned above, we see examples of how new literacy practices are developing in schools but also the constraints traditional schooling presents to the development of such practices. A basic requirement of schools today is, therefore, to deal with changes that have resulted in part from technological developments.

Many schools have problems in developing learning strategies supporting remixing practices because the structure of the school day, teacher competencies, examination systems, and so forth, do not take into consideration developments in the ways young people use new technologies. The institutional practice of schooling is thereby challenged by remixing as a literacy practice, entailing students taking a more active role in developing knowledge. A key question in these developments involves interrogating what we have traditionally meant by the distinction formal versus informal learning. From the perspective of young people learning takes place in many different contexts, taking experiences from one context over to another. Remixing is an activity that cuts across such educational conceptions.

Conclusion

Norway is an example of a country that has taken the step from looking at digital media as an object that has to be implemented in school settings to asking questions about the real implications this has for how we conceive learning and literacy. Digital literacy has now been written into its national curriculum as one of five key competences. The consequent challenge is to discover how to make this work in educational practice and to let the experiences of young people outside schools inform the constant redefinition of the social practice of schooling. I believe remixing represents an area that we have to take seriously for future developments in schooling.

Note

1. In Norwegian there is no word for the English term "literacy." Traditionally it has been translated with alphabetization, but during the last 20 years the term "competence" has been used instead, with a broader conception of reading and writing.

References

Ades, D. (1976). *Photomontage*. London: Thames and Hudson.

Alvermann, D. (Ed.). (2002). *Adolescents and literacies in a digital world*. New York: Peter Lang.

Barthes, R. (1978). *Image, music, text*. London: Hill and Wang.

Bolter, J.D., & R. Grusin. (1999). *Remediation: Understanding new media*. Cambridge, MA: The MIT Press.

Bordwell, D. (2005). *The cinema of Eisenstein*. New York: Routledge.

Buckingham, D. (2003). *Media education. Literacy, learning and contemporary culture*. Cambridge, UK: Polity Press.

———. (2007). Defining digital literacy: What do young people need to know about digital media? *Nordic Journal of Digital Literacy*, 4, 263–276. Oslo: University Press.

Buckingham, D., & R. Willett (Eds.). (2006). *Digital generations: Children, young people, and new media*. Mahwah, NJ: Lawrence Erlbaum.

Castells, M. (1996). *The rise of the network society, the information age: Economy, society and culture*. Vol. I. Oxford: Blackwell.

Cope, B., & Kalantzis, M. (Eds.). (2000). *Multiliteracies: Literacy learning and the design of social futures*. London: Routledge.

Diakopoulos, N. (2005). Remix Culture: Mixing Up Authorship. Retrieved November 20, 2007, from http://www-static.cc.gatech.edu/~nad/

Erstad, O. (2005). Expanding possibilities: Project work using ICT. *Human Technology: An Interdisciplinary Journal on Humans in ICT Environments*, 1(2), 109–264. Retrieved November 20, 2007, from www.humantechnology.jyu.fi

Erstad, O., Klovstad, V., Kristiansen, T. & Soby, M. (2005): *ITU Monitor 2005: Digital kompetanse i skolen* [ITU Monitor 2005. Digital literacy in the school]. Oslo: The Norwegian University Press.

Erstad, O., Gilje, Ø., & de Lange, T. (2007). Re-mixing multimodal resources: Multiliteracies and digital production in Norwegian media education. *Journal of Learning, Media and*

Technology. Special Issue: Media education goes digital, D. Buckingham & S. Bragg (Eds.). London: Taylor & Francis. 32 (2), pp. 183–199.

Erstad, O. & Silseth, K. (2008) Agency in digital storytelling: Challenging the educational context. In K. Lundby (Ed.), *Digital storytelling, mediatized stories: Self-representations in new media* . London: Peter Lang.

ETS (2002). *Digital transformation: A framework for ICT literacy*. Princeton, NJ: Educational Testing Service.

European Commission. (2004). *Key Competences for Lifelong Learning: a European Reference Framework*. Directorate-General for Education and Culture. Retrieved November 20, 2007, from http://europa.eu.int/comm/education/policies/2010/doc/basicframe.pdf

Hebdige, D. (1979/1985). *Subculture: The meaning of style*. London: Methuen.

Heim, J., & Brandtzæg, P.B. (2007). Patterns of Media Usage and the Non-professional Users. Paper presented at the Computer/Human Interaction Conference, April 28—May 3, San Diego, US.

Hull, G., & Greeno, J. (2006). Identity and agency in nonschool and school worlds. In Z. Bekerman, N. Burbules, & D. Silberman-Keller (Eds.), *Learning in places: The informal education reader* (pp. 77–98). New York: Peter Lang.

Ito, M. (2006). Japanese media mixes and amateur cultural exchange. In D. Buckingham & R. Willett (Eds.). *Digital generations: Children, young people, and new media* (pp. 49–66). Mahwah, NJ: Lawrence Erlbaum.

Jenkins, H. (2006). *Convergence Culture. Where old and new media collide*. New York: New York University Press.

Kellner, D. (2002). Technological revolution, multiple literacies, and the restructuring of education. In I. Snyder (Ed.), *Silicon literacies. Communication, innovation and education in the electronic age* (pp. 154–169). London: Routledge.

Kittler, F. (1990). *Discourse networks 1800/1900*. Stanford, CA: Stanford University Press.

Kozma, R. B. (Ed.). (2003). *Technology, innovation, and educational change: A global perspective*. Eugene, OR: International Society for the Evaluation of Educational Achievement.

Kress, G. (2003). *Literacy in the new media age*. London: Routledge.

Landow, G. (2006): *Hypertext 3.0: Critical theory and new media in an era of globalization*. Baltimore: The Johns Hopkins University Press.

Lankshear, C., & Knobel, M. (2006). *New literacies, everyday practices and classroom learning*. Berkshire, UK: Open University Press.

Lemke, J. L. (1998). Metamedia literacy: Transforming meanings and media. In D. Reinking, M., McKenna, L. D. Labbo, & R. D. Kieffer (Ed.), *Handbook of literacy and technology: Transformations in a post-typographic world*. Mahwah, NJ: Lawrence Erlbaum Associates.

Lessig, L. (2005). *Free culture: The nature and future of creativity*. New York: Penguin.

Livingstone, S. (2002). *Young people and new media*. London: Sage Publications.

Manovich, L. (2007). What comes after remix? Retrieved November 20, 2007, from http://remixtheory.net/?p=169

Olson, D.R., & Cole, M. (2006). *Technology, literacy, and the evolution of society: Implications of the work of Jack Goody*. Mahwah, NJ: Lawrence Erlbaum.

Pahl, K. & Rowsell, J. (2005). *Literacy and education. Understanding the new literacy studies in the classroom*. Thousand Oaks, CA: Sage.

Peppler, K.A., & Kafai, Y.B. (2007). From SuperGoo to Scratch: Exploring creative digital media production in informal learning. *Journal Learning, Media and Technology. Special Issue: Media Education Goes Digital,* guest editors: D. Buckingham & S. Bragg. London: Taylor & Francis. *32(2),* pp. 149–166.

Rasmussen, I. (2005). *Project work and ICT. Studying learning as participation trajectories.* Dissertation, Faculty of Education, University of Oslo, Norway.

Rassool, N. (1999). *Literacy for sustainable development in the age of information.* Clevedon: Multilingual Matters Ltd.

Reinking, D., McKenna, M.C., Labbo, L.D., & R.D. Kieffer (Eds.). (1998). *Handbook of literacy and technology. Transformations in a post-typographic world.* Mahwah, NJ: Lawrence Erlbaum.

Scardamalia, M., & Bereiter, C. (2006). Knowledge building: Theory, pedagogy, and technology. In R. Keith Sawyer (Ed.), *The Cambridge handbook of the learning sciences* (pp. 97–115). Cambridge: Cambridge University Press.

Scribner, S., & Cole, M. (1981). *The psychology of literacy.* Cambridge, MA: Harvard University Press.

Snyder, I. (Ed.). (2002). *Silicon literacies. Communication, innovation and education in the electronic age.* London: Routledge.

Säljö, R. (1999). Learning as the use of tools: a sociocultural perspective on the human-technology link. In K. Littleton & P. Light (Eds.), *Learning with computers: Analysing productive interaction (*pp. 144–161). London: Routledge.

Tyner, K. (1998). *Literacy in a digital world: Teaching and learning in the age of information.* Mahwah, NJ: Lawrence Erlbaum.

Wertsch, J. (1998). *Mind as action.* New York: Oxford University Press.

CHAPTER NINE

Crossing Boundaries

Digital Literacy in Enterprises

LILIA EFIMOVA AND JONATHAN GRUDIN

Introduction

New technologies are often embraced by the young, who can soak up the capabilities and nuances much as they absorb language, culture, and traditional literacies. An earlier generation of students picked up text editing and email, and subsequently brought their skills and knowledge of uses and effective practices into workplaces. Communication technologies particularly appeal to young people, whose primary focus is often on building social networks and learning about the world from their peers—and, occasionally, from elders, especially those who learn to speak the "same language."

Cost can be a barrier to accessing some communication technologies for young people, but they are motivated to overcome such barriers and find ways of accessing communication technologies by sharing systems at schools or libraries, hanging out with more fortunate friends and siblings, and pressuring parents. Mobile phone use is progressively reaching younger and younger age groups, even in less prosperous regions. Today, young people lead in the use of text messaging, instant messaging, blogging, and social networking and communication software use in general (e.g., using MySpace, Facebook, YouTube, Skype, and so on). In doing so, they acquire facility with features of these technologies, and understand challenges the technologies can and cannot address, problems that use might engender, and social conventions that govern effective use.

As students leave school or university and move into workplaces and other organizations, they carry these skills and knowledge with them. They can see where these technologies can address problems or improve efficiency. However, most of their new, albeit typically older, colleagues are unfamiliar with the technologies and skeptical of their proposed uses. The older generation has social networks in place and tends to be less focused on informal communication, so will more slowly try new technologies and learn new tricks than do young co-workers. So the spread of digital literacy into enterprises is often resisted. It comes slowly. But it comes.

In this chapter we first illustrate this phenomenon by examining the parallel between the adoption of email and the adoption of instant messaging twenty years later. We then turn to a third example and our main focus, a case study of blogging in a large high-tech company. Weblogs typify social software empowered by web-based visibility which may move into enterprises more rapidly than previous technologies. Infrastructures are already in place for enterprise-based blogging; the software is relatively "lightweight" and inexpensive, and competitive drive for overall efficiency—which, we argue, can be enhanced by workplace blogging—is stronger in global markets.

Digital Literacies and Enterprises: A Parallel Between Email and Messaging

Email in the 1980s

Today, we take email for granted. Indeed, it is a mission-critical tool for most enterprises in the developed world and increasingly significant across the planet. Only twenty years ago, however, this was not true. In the 1980s, email was regarded with suspicion by management, even in high tech firms. Popular with university students, yes, but would it distract employees from productive work? Even in the 1990s some researchers thought organizations would abandon email once they measured productivity losses (e.g., Pickering & King, 1992).

Indeed, one of us experienced this first-hand. Employed in 1983 at a high-tech firm that had deployed email but did not use it, he was told that email was a way students wasted time, and that to get information from people elsewhere in the organization, he should learn to write formal memos and send them through the management chain.

In the 1980s, disk space was too expensive to use to save email after it was read. University students learned to use it for informal exchanges and quick questions and answers, based on a conversational model, ignoring typos, sprinkling in strings of exclamation points, question marks, or capitalized words to mimic spoken emphasis. The older generation associated typing with formality and permanent records. Some older folk who began using email treated it like written correspondence, starting each message with "Dear . . ." and closing with "sincerely" or "yours truly."

Practices, attitudes, and technology changed. More young people entered the work force, where informal social interaction and quick business exchanges are important, after all, and found email a natural medium. The cost of acquiring and maintaining an email system soon came down, and disk sizes increased, so email could more often be saved. In the 1990s, email attachment functions became widely available, allowing the sharing of formal documents, spreadsheets, and slide decks much favored by managers. Accordingly, email became acceptable for more formal correspondence. Moreover, with greater likelihood of saving emails for future review, senders felt more accountable for content and form. Strings of exclamation points disappeared; tools came along to help fix spelling and grammar. The left-hand column of Table 9.1 identifies some of the key characteristics of email in 1983. Some might surprise people today. For example, the lack of naming conventions for email addresses or the absence of online directories meant one could mainly just email friends who had formally established email accounts.

The informality and ephemeral nature of email during the early 1980s was a key obstacle to enterprise adoption; managers just didn't see any value in it. Nevertheless, individual contributors to an enterprise often have a high need for quick questions and answers and informal discussion. These individuals are typically collocated in groups. On one hand, these groups often have fewer meetings than was the norm previously, so the asynchronous interruptions of email are manageable between meetings. On the other hand, managers and executives rely more heavily on exchanging structured information, spend more time in meetings where interruptions are problematic, and tend to regard informal "chatting" by their employees with suspicion. A balance must be struck, but informal communication remains important.

Table 9.1. Characteristics of Email in 1983 and in 2008.

Email in 1983	Email in 2008
• used mostly by students • access limited to friends • email clients not interoperable • conversations ephemeral • attachments not supported • chosen for informality • organizational distrust: chit-chat? ROI?	• used by everyone • accessible to everyone • complete interoperability • conversations saved • attachments a major feature of email • has become formal • email is mission-critical to the workplace

Illustrating this point, in 1983, the head of Xerox PARC said that a company executive had confessed that embarrassment about his spelling kept him from using email and asked whether PARC could develop an automatic email spelling corrector. "We could," John Seely Brown reported telling him, "but we won't. Instead, we'll build a spelling de-corrector!" He argued that email was and should remain an informal communication medium. A tool that inserted random spelling errors would help keep it informal.

Messaging in the 2000s

Email did not remain informal, of course. The spelling de-corrector wasn't invented—but instant messaging was! A need persisted for casual channels for casual Q&A, unpolished ideas, exaggeration, sloppiness, letting off steam, gossiping and flirting, without fear that unguarded remarks would return to haunt the sender. Young people lauded instant or text messaging for "not being formal, like email," which is what the latter had become. Just as email was quicker than formal memos, instant messaging (IM) is quicker than email. In the early days of messaging, no one worried about spelling, attachments were rarely supported, one mainly reached buddies, and messages were rarely saved.

Like email before it, messaging was regarded with managerial suspicion. From Gartner, the major consulting firm, came the following in 2005: "Prediction: IM misuse will threaten user productivity . . . IM misuse and overload has the potential to be worse than email overload . . . Enterprises run the risk of turning unmanaged, unsanctioned consumer IM into unmanaged, sanctioned EIM" (Grudin, Tallarico, & Counts, 2005, no page). This, too, is changing, however, and more rapidly due in part to the comparatively low cost of messaging. We find that with high demand for archiving IM conversations and for sending attachments, formality is likely to increase along with managerial acceptance (see Table 9.2). Digital literacy will again move from student and consumer spaces into enterprises.

Table 9.2. Characteristics of Instant Messaging in 2003 and in the Near Future.

Instant Messaging in 2003	Instant Messaging in the Near Future
• used mostly by students • access limited to "buddies" • im clients not interoperable • conversations ephemeral • attachments rarely supported • chosen for informality • organizational distrust: chit-chat? ROI?	• use spreading rapidly • pressure to remove limits • presure for im client interoperability • recording is more common • attachment capability sought • recording → increased formality • will become essential in the workplace

A Case Study of Weblog Use

Next we present a study of emergent blogging practices in a corporate setting. We attended meetings, read email, documents, and weblogs, and interviewed 38 people—bloggers, infrastructure administrators, attorneys, public relations specialists, and executives. We found an experimental, rapidly evolving terrain marked by growing sophistication about balancing personal, team, and corporate incentives and issues.

Weblogs are used by millions of people. Research is being published on genres of use (Herring, Scheidt, Bonus, & Wright, 2004), motivations and expectations (Efimova, 2004; Nardi, Schiano, & Gumbrecht, 2004; Viégas, 2005), and other aspects of what the December 2004 special issue of *Communications of the ACM* titled "the blogosphere." Most weblogs are written by individuals for friends and family or to inform the public about personal views and observations. These range from diary-style student weblogs to "A-list" weblogs maintained by political candidates, journalists, pundits and other prominent people on a range of topics. Most bloggers are in their late teens and early twenties. If history is a guide, they will carry skills and knowledge about weblog capabilities into workplaces. In an earlier era, students adopted line and text editors, forerunners of word processors, at a time when keyboard use was considered blue-collar and was avoided by knowledge workers and managers. Employees who had adopted email as students found that even high tech workplaces were skeptical about its value (Perin, 1991). More recently, instant messaging followed a remarkably similar path of student adoption, corporate suspicion, and ultimately growing acceptance (Lovejoy & Grudin, 2003).

Corporate adoption was slow for word processing and email. Word processing did not become widespread until a generation with keyboard skills arrived. In most organizations, email required significant new infrastructure—hardware, software, and administrative support. Today's emerging technologies will have an easier time. IM clients are easily downloaded. Free or inexpensive web-based weblog technology is readily available. Costs for organizational hosting remain but are substantially lower than in the past.

How quickly are corporate or employee weblogs likely to take hold? Elsewhere (Efimova and Grudin, 2007), we identify and describe a recent projection that identifies corporate weblogs as rapidly approaching mature use. According to this projection corporate blogging was expected to enter productive use as early as 2007–2008.

Employee Weblogs

Some people define a blog as online writing designed for a wide public audience. We use the term more inclusively—if an author considers it to be a blog then that suffices—though our focus here is just on blogs that touch on worklife (Efimova, 2004). We use the term "employee weblog" instead of "corporate blogging" because the latter suggests action that is authorized, acknowledged, or formally associated with an organization.

Some weblogs focus on personal life and mention work in passing; others focus on work experiences and say little or nothing about personal life. Reflections on work may be general or specific to an author's project or group. Intended audiences vary: friends, acquaintances, fellow employees, people interested in general aspects of work life such as those found in a novel, colleagues or fellow professionals, customers or partners of the author or employer, or external media interested in the organization, to name a few. Because weblogs are often highly visible, easily accessed, and indexed by search engines, their use by employees raises issues for teams and organizations. With a few keystrokes, information traverses the wall separating an organization from the outside world. Planning and social convention go into erecting and maintaining such walls; it can be unsettling to have them so easily crossed. Although in principle not unlike sending an email attachment or newsgroup post, the instantaneous, wide visibility can feel qualitatively different, amplified by ripple effects or information epidemics created by blogger networks (Adar & Adamic, 2005). The effect is most strongly felt when readers can identify an author or the author's organization.

For a large company, weblogs present an untested middle ground between public relations handled by professional staff and the usually inconsequential employee discussions of work with family and friends. Even when pitched mainly to family and friends, weblog posts may be indexed by search engines and delivered by watchlists minutes after being written. This is complicated by the fact that people are not always careful with what they post. In April 2003, one of us chanced upon a weblog, public—although written mainly for friends—in which a colleague described actions that were clear grounds for job termination. In widely publicized events, a Google employee was fired for discussing everyday life at work (Cone, 2005), a Microsoft contractor for posting a photograph of a company site (Bishop, 2003), and employees at Delta Airlines, ESPN, and Waterstone Books for blog content. (Searching on "fired for blog" yields hundreds of hits.)

At the same time, employee blogging is starting to be seen as a potentially useful communication channel. Zerfaß (2005, discussed in Jürch & Stobbe, 2005) describes eight functions of corporate blogs. One is pure public relations, two is to deliver internal communication—knowledge transfer and contract negotiation—and five focuses on market communication: product blogs, service blogs, customer relationship blogs, crisis blogs, and CEO blogs (which we broaden to executive blogs, e.g., Dudley, 2004, and which can also serve an internal communication function). Accounts of employees blogging openly about work appear regularly (see, for example, Edward Cone's "Rise of the Blog," 2005). Weblog authors in major technology companies can be found by searching for "(company name) bloggers," where the company name is Amazon, Google, IBM, Microsoft, Sun, and so on. The resulting lists are neither official nor comprehensive, but they reveal that employee blogging is widespread. The growing familiarity of young people with the form and analyses of its potential motivate a look at early adopter organizations, teams, and individuals (Grudin, 2006).

How do weblog authors balance writing about work and personal life? How do they react to feedback and comments from inside and outside their organizations? How does management deal with shifting external perceptions of the company and its employees? Weblogs could affect legal, public relations, and human resources policies and practices. What are the risks, the possible benefits? The millions of young people entering organizations will know how to work more effectively and efficiently by applying new technologies, communication skill, and have the ability to create engaging digital multimedia, much as their predecessors made email and word processing mission-critical tools. Studies of early adoption can contribute to designing these technologies for organizational settings and to guiding organizations in their use.

Study Goals and Method

Our study site was a single site: Microsoft. Reports indicate that even within high-tech companies, weblog use varies considerably. This may reflect differences in size, geographic dispersion, corporate culture, or happenstance. Nevertheless, many individual incentives and experiences, and organizational opportunities and sensitivities, are likely to be common.

The second author, Jonathan Grudin—one of about 60,000 Microsoft employees—has created weblogs but was not part of the company's active weblog culture. The first author, Lilia Efimova—a relatively well-known blogger—visited Jonathan's place of employment to work on the study. We set out to explore where, how, and why employees blog; how personal the writing is in work-related weblogs; what happens when blogging becomes a formal work objective; perceptions of the personal and business impacts of blogging; and possible steps to make blogging more effective.

Over ten weeks (July to September, 2005) we browsed and read employee weblogs, followed weblog email distribution lists, attended meetings organized by others to discuss weblog issues, read documentation covering weblog guidelines and policies, and interviewed 38 people in the organization, most in-person for an hour or more, and some by phone. We had access to data from internal surveys that covered weblog awareness, attitudes, and behavior. We first interviewed employees who had supported, promoted, and authored weblogs, gathering relevant history and identifying significant groups and roles. These interviewees included: active bloggers, infrastructure support (e.g., those managing servers) and policy-makers (e.g., attorneys). These people suggested other interview candidates; yet others we found by exploring employee weblogs and contacting authors whose weblogs complemented those in our sample. These included well-known and less well-known bloggers, employees in different roles or located in different countries, those with diverse blogging styles (strictly work-related, mixing work and personal, product-oriented blogs, internal weblogs that could be classified as "project weblogs"—see Udell, 2005—and non-English weblogs). Table 9.3 provides an overview of our sample.

	Total	Male	Female	Infrastructure or policy
Bloggers	34	29	5	7
Non-bloggers	4	3	1	4
Total	38	32	6	11

Table 9.3. Interview Respondents.

Semi-structured interview questions addressed history, perceptions of blogging in the organization, and personal practices emphasizing respondents' knowledge of or involvement in organization-wide blogging processes. Specific questions about events or blog content were based on insights gained from reading their blogs. Over time some emphases shifted. For example, the discovery of heavy product weblog activity led to more exploration of that particular focus.

Virtually everyone we approached agreed to be interviewed and engaged enthusiastically with the research. This may partly reflect the verbal, discursive nature of blogging, but many of our questions clearly resonated with people's perceptions and reflections on this rapidly evolving communication medium.

Results

Our primary focus was on weblog authoring and the authors' views of the readership. After describing the evolution of perceptions and policies around blogging, we present an overview of weblog infrastructure. Then work-related uses of weblogs and their implications are discussed, followed by a case of product weblogs. These types of blogs are an active form that illustrates some of the issues and patterns we observed. Finally we discuss personal choices that shape blogging and close with a discussion of the implications of our findings for enterprise contexts.

Evolving Perceptions and Policies

The first Microsoft bloggers were university students with externally hosted weblogs who were hired as interns or employees, starting in 2000 and 2001. Their weblogs attracted little attention. By mid-2002 employees were manually hosting weblogs on company machines and arguing for externally visible weblogs. An internal weblog server, maintained through voluntary efforts, hosted a few dozen weblogs by the end of the year. Late in 2002 a list of employee weblogs, including some hosted externally, was published by someone outside the organization (Mary Jo Foley in *Microsoft Watch*). This helped create a sense of a community engaged in externally visible blogging. The attention led to internal meetings and reflection.

Internal servers are necessary for internally facing weblogs accessible on an intranet, but not for externally facing weblogs, which can be hosted on any server. However, by dedicating servers to host external weblogs, a company can facilitate, promote, and possibly monitor activity. A successful grassroots push by passionate employees for such servers gave rise to issues of ownership and appropriate behavior. By mid-2003, a server hosting externally visible weblogs was operating. Because some managers perceived a benefit in using weblogs to communicate with customers, this server had formal budget support.

The wisdom of letting employees blog was actively debated by those aware of these efforts. Indeed, many early bloggers within the company felt that legal and public relations representatives wanted to shut them down. In an open internal panel discussion in June 2003, a legal representative benignly encouraged bloggers uncertain about the wisdom of publishing particular content to seek guidance. Four months later, however, a contractor was dismissed for what many considered a relatively minor disclosure in a blog. Many in the weblog community had made similar disclosures, so there was great concern. The resulting discussions among bloggers, human resources, legal, and public relations were seen as producing healthy mutual education and clarification of policy.

We interviewed two senior attorneys charged with considering weblog activity. They noted that long-standing policies covering email and newsgroup posting applied to weblogs. They recounted examples of employees saying unwise things in public weblogs—often humorous in retrospect—but noted that similar incidents occur in other media. The attorneys appreciated that employee weblogs enabled the company to very rapidly counter misinformation in press coverage and had even assigned a law student intern to research the benefits and drawbacks of initiating a public weblog focused on legal issues. The attorneys noted that Bill Gates and Steve Ballmer had spoken positively in public about weblogs. A senior vice president began blogging in May 2003. Not all executives showed the same level of enthusiasm, but by the summer of 2005 the climate had shifted. The attorneys suggested that an event like that of two years earlier would not lead to dismissal today. Guidelines for weblog practice had occasionally been circulated. People were sensitive over how to characterize them. Repeatedly we were told "the policy is that there is no policy," or "the policy is 'be smart.'" Some worried that even these would lead to the charge "You have instituted a blogging policy!" The attorneys backed a mild "be smart" policy, while noting pointedly that other policies cover the disclosure of proprietary information applied in this medium.

Public relations staff are potentially affected by blogs in two ways: weblogs can create problems for them to handle, and weblog success could undermine their role. Indeed, we were told that some managers were considering diverting some publicity funds into hiring a blogger. Blogger concern that Microsoft's Public Relations group would be antagonistic was not unreasonable. It was company policy to bring in Public Relations when interfacing with external media. This was not applied to online newsgroup participation, but weblogs are more likely to be noticed by external media and disrupt carefully timed media campaigns. In an interview with us, a senior manager in the public relations group demonstrated a very sophisticated understanding of weblogs. She saw them as a channel that would affect but not supplant other channels, bringing benefits and risks. Her job was to understand and shape effective practices in a shifting terrain. Complex issues of control would no doubt arise, but she saw that the clock would not be turned back.

We interviewed two vice presidents of product development. One, unabashedly enthusiastic, had hired a well-known blogger. He argued that the company had much to gain from being seen as open and transparent. The other vice president was skeptical. He had concerns about self-appointed spokespeople for a project or for the entire company. Although perceived to be antagonistic by bloggers with whom we spoke, during the course of our study he initiated a blog himself, with a focus on recruiting. He also supported the initiation of a product blog in his organization. He realized that weblogs must be credible and relatively informal but stressed strategic planning, with careful consideration of consequences, including possible effects on team members should one person become well known based on the group's work.

Weblog Infrastructure

A complete overview of employee blogging proved to be impossible. Company-hosted server weblogs were visible but were only a fraction of the activity. The server administrators estimated two to three thousand bloggers in the company, but an internal survey put the number at over seven thousand. This imprecision is understandable: employees are not obliged to use official company servers, to report that they are starting a weblog, or to identify themselves or their affiliation. Drawing on data from different sources, we were content with identifying major weblog categories and estimating their numbers when possible.

At the time of the study, *internally hosted weblogs* at Microsoft include approximately 800 on a central server with an unknown number of self-hosted weblogs on other intranet servers. At Microsoft, external customer-oriented weblogs are perceived to be the principal value of the medium, with skepticism toward internal weblogs—"there is not a clear business purpose for it"—which are therefore not supported formally, the server being maintained by volunteers and intermittently down (a new server was donated by unhappy users to solve this problem). An index of internal weblogs is consulted in intranet searches, making it a good tool "to add to that index." But they are only accessible when an employee is on the corporate network, which obstructs access to one's own internal weblog while mobile.

Weblogs hosted on *external official servers* are publicly accessible but can only be created by employees. Servers run by two corporate groups hosted over 2000 weblogs; several regional servers host weblogs authored in local languages, creating local faces for an international company.

Company-supported external servers include those available to the general public but supported by Microsoft, specifically weblogs that are part of the company community initiatives and the MSN Spaces consumer blogging platform. These are intended to provide blogging space for non-employees, but nothing prevents employees from using them.

Finally, *other external servers*—public blogging platforms and self-hosted servers—have no Microsoft connection or dependencies and are difficult to identify for research purposes.

Internally, there are two email lists dedicated to blogging issues and some document repositories, although the latter originated in different departments and are not easily found (some respondents were unaware of them).

Work-related Uses of Weblogs

From the interviews came three broad categories of weblog use: direct communication with others, showing a human side of the company, and documenting and organizing work. Many described blogging as a way to *share their passion for their work and to communicate directly with others inside and outside the organization*. Often, people who design and develop a product have unique information but are separated from customers and users by intermediaries in sales, marketing, and field support, and by the time to reach market. Writing formal articles that could be published on the company web site was not appealing to our informants due to the time and rounds of reviewing required to publish via official channels, and the lack of visibility or feedback associated with such materials. In contrast, a weblog is an easy way to provide information, share tips, and engage in direct interaction with peers outside the organization or with consumers of one's work. One respondent noted, "we were trying to ship something and [in my role] I have no external exposure to people ... so [starting a weblog] was partly to talk about it with outsiders." The visibility of blogs via search engines ensures that posts are relatively easy to discover. Another respondent received permission to publish internal FAQ materials in his weblog to benefit external readers.

Most bloggers found it gratifying to inform or help others, to learn about the destiny of their work in the "real world," or to become visible as an expert in a specific area via their blogs. Company encouragement to interact with customers and engage with communities provided a supportive atmosphere and eliminated potential barriers but did not seem to directly induce blogging. As one person put it, "blogging doesn't come out of fear, it's about passion." As employees of a company that can seem impersonal to those outside, many described a desire to *show the human side of the company* (see Lovejoy & Grudin, 2003, for an indication that weblogs can be effective in this respect). They wanted to demonstrate that people in the organization care and are passionate about their work. They could recount stories behind products to help people understand why particular choices were made and share details of daily routines to give outsiders a sense of the context of their work. One respondent said, "I'm tired of being called evil." Bloggers also felt they could respond in crises with greater credibility based on a history of objectively sharing useful information. Where the company's primary language is not spoken, this was particularly emphasized and some country-specific blog servers have been set up. Writing in local languages enables greater connection with those communities.

Humanizing the company in the eyes of potential employees was also emphasized. Three informants (two HR employees, one vice president) consciously crafted weblogs for recruiting. Their weblogs told everyday work stories for different roles in the company, provided insight into selection or promotion procedures, and shared tips and tricks. These authors felt the weblogs had measurable impact on recruitment. Other people reported new hires who had applied to a group after reading a group member's weblog.

Some employees used a weblog both to communicate with others and as a space to *document and organize their work* or draft ideas. Several described their weblog as a personal archive enhanced by feedback from readers; "either I could have written that down as an internal note and just kept that or now it's out there on internet, so I can find it more easily and also get hints from folks." A few people mentioned that they enjoy writing, and two had aspirations to write a book based on weblog entries. Several internal weblogs, including one by a team, were used to document and share work in progress with others.

That weblog content can have long-term value for an individual is seen in this comment on future access to an internally hosted, externally visible weblog; "if I leave the company they say it could be archived, but you will not be able to update it [. . .] if they said they would delete it, I'd be thinking why am I blogging here and not externally . . . and grab my old content." Bloggers who do not mention documentation as a major motivation sometimes use old entries in drafting more formal documents, or save time answering a frequently asked question by sending a link to a blog entry. Several people indicated that they could avoid "spamming" others with experiences and ideas by placing them in an easily accessible weblog post.

Implications: Finding and Being Found

In employee weblogs, ideas that were previously unarticulated or hidden in personal archives become visible, interlinked, and searchable. Collectively, this produces a wealth of information about products, practices, tips and tricks. Many respondents reported time saved by blogging: re-using entries, quickly helping others or learning, getting answers to questions, receiving feedback on ideas, finding people inside or outside the company with similar interests or needs. A few bloggers mentioned that posting to their external weblog helped them connect serendipitously to a person or relevant information inside the organization. One noted that an idea posted to a weblog resulted in a prototype developed in another part of the organization. He wrote, "I've never met Lee or had any agreements with anyone that he would do this. Nor would I ever have been able to send mail to the right group of interested people that might be able to spend the time building a prototype. I simply blogged my idea, the idea found the right people, and we've made a bunch of progress that will help ensure the right feature is delivered to our users."

A weblog also gives visibility to its author, whose expertise can be exposed beyond his nearest circle of colleagues. Our informants told us about invitations to publish articles or speak at events as a result of blogging. Several reported that their job responsibilities evolved as their interests were exposed: "[After reading my weblog my manager said] if you are so externally focused, you can be our community lead . . . now I'm a community lead . . . I enjoy it." Some bloggers noted that being recognized as an expert gave them greater confidence in their career prospects. Indeed, it seems that externally visible blogging provides publicity that many work roles and positions would not normally entail. Some bloggers acquired more negotiating power or job security as people realized that making them uncomfortable or dismissing them could have repercussions with customers or partners. Blogging externally was also seen as a way of helping to accelerate internal change: suggestions made in public may get more attention than those delivered internally. In addition, customer feedback can confirm ideas, giving a proposal more validity.

Of course, these power shifts can lead to tension, so visibility can be a mixed blessing. Some bloggers dislike the limelight and experience or worry about tensions within their teams when readers credit them for a team effort: "You are not trying to expose yourself or to be a star." Also, becoming a contact point for customers raises expectations for blog coverage and the blogger becomes a focal point for questions and suggestions. Bloggers with large audiences complained of email overload and discussed preventive measures. Some felt they were doing other people's jobs on top of their own.

The Case of Product Blogs

This section focuses on a specific type of employee weblog, strongly associated with a specific product in the eyes of readers. We distinguish two types of product weblogs: those intentionally focused on a product from the beginning, and emergent, de facto product weblogs. A product weblog can provide a product team with an unmediated way to engage customers, to learn about their experiences, and to reveal human faces behind the product.

Intentional product weblogs focus less on individual personalities; they provide informal views and timely information behind specific products and engage with customers who use them. They supplement rather than replace formal PR and marketing, providing stories about the decisions that shaped the product, time-sensitive information that would take too long to publish through formal channels, and tips and tricks. For readers, such a product blog can be a single place to get news about a product and to communicate directly with people behind it. It feels more official than a personal employee weblog. This can yield a bigger readership and greater impact but has risks as well. With a product blog written by a team, more is at stake: readers' expectations about content quality and regularity are higher than for a personal blog. The authority of a product blog increases the potential impact of a mistake; if the weblog creates news it can engender a PR crisis. One respondent noted that PR specialists responsible for a product asked his team not to blog on Fridays: "you're gonna impact their lives [if an emergency arises over a weekend]."

Most product weblogs authored by our respondents were team endeavors, although one person might lead the effort and exhort other team members to blog. In all but one team product blogs, entries appear with an author name, showing the personality and style of each team member and ensuring personal accountability. Some respondents considered this a critical aspect of team blogs and complained that their weblog technology did not support including author as metadata for searching or filtering. Thus, given expectations of a topical focus and stronger ties with an official product or company image, product weblogs generally include some constraints on content or style. Personal entries were considered less appropriate in this context, but no one indicated that was a strict rule; in fact, one noted, "we didn't get killed for personal stuff [on the product blog]."

Every intentional product weblog we saw had an editorial process. The specifics varied greatly. Some product teams collected and reviewed ideas or drafts via a group mailing list, document server, or in meetings. In some cases agreement of all team members, including marketing representatives, was required to post. In others, reviews were only used to get opinions about questionable content. Reviews were variously used to ensure regularity of postings, obtain consensus between personal opinions and overall team perspectives, and to block information with high risk of misinterpretation or misuse by the external audience. Editorial processes can reduce risk and increase uniformity, but of course they can have negative impacts as well. Review and negotiation take time—in some cases up to a couple weeks—which reduces the immediacy that is integral to blogging, making it more like other forms of corporate web publishing. Review can reduce the informality and the motivation of individual contributors; one respondent mentioned the "pain of being edited by your colleagues." Some contributors to a product blog write even more about the product in their own work-related weblogs, where they have more freedom and flexibility. One noted, "the problem with team blogs: because everyone can contribute, doesn't mean they will." On the other hand, blogging together lowers the pressure on any one person to provide interesting material regularly and reduces the time required of a solo weblog author; some team bloggers definitely appreciated that. However, believing that group posting and an editorial process can kill the personality and immediacy that appeal to potential readers, some bloggers are extremely critical of team product blogs; "my feeling is that people don't like team blogs as much as personal blogs . . . [Other company] blog feels like a press-release." It is unclear why team product blogs are perceived that way. It may be due less to the group authorship per se than to the editorial process it often implies, and to self-editing of style and content to avoid possible negative impact.

De Facto Product Weblog

An alternative form of blogging that has similarly strong ties to a product is a *de facto product weblog*. De facto product weblogs are created as personal weblogs, often written outside job responsibilities, and not as the focal point for product information. The product focus emerges as their authors post on themes they are knowledgeable and passionate about. Their authors feel less pressure to conform to product group norms or official PR initiatives. However, some become strongly affiliated with a particular product or initiative in the eyes of external readers, giving rise to the same risks and potential business benefits as intentional product weblogs. Management may see a de facto product weblog as a potential communication channel to reach customers or an external community. One person in a public relations role (a blogger himself) described a complex situation that arose with a de facto product blogger: "we wanted to get into the community and asked him to post something . . . ask him to post our press-releases, so enthusiast groups can get them . . . media alerts . . . what's happening officially . . . it is not the best thing for him or us [. . .] don't want him to be the official spokesperson . . . for him it's also putting out official information and he feels less free to comment on that . . . also some of his readers would suspect that his weblog is written by a corporate guy—'you are not one of us, but one of them.'" He then described his plan to start a 'proper' weblog for the product that would provide a more person-independent, objective space for informal communication and engagement with customers.

Our data show that another potential problem arises when an author of a de facto product weblog moves to another position in the company, leaving old interests behind and wishing to shift weblog posts to describe new job challenges. For the audience it could be an abrupt loss of a space to receive information and to engage with others, and this could have negative consequences for the company. To transfer the weblog to another author wouldn't work here: The weblog wasn't intentionally created to have a product-focused purpose; it was centered on personal interests and strongly tied to its author.

Personal Choices

Blogging is still an area of experimentation at Microsoft and it is generally up to a given individual to decide if, when, why and how to blog. We identified several choices a blogger who works for the company had to make.

Starting a Weblog

Most people we spoke with began on their own initiative, with little prior discussion. "I asked only my direct manager and it was on purpose: I knew if I [were to] ring my manager's manager or manager of my manager's manager it would become impossible." Many bloggers cited experimentation, examples set by colleagues or pressure from others as reasons for starting a weblog. Almost everyone mentioned a work-related rationale for blogging. Personal reasons for starting to blog were central in the case of strictly personal weblogs—"it proved to be a good communication tool with my friends"—and also appear in weblogs that include work-related content. With the latter, personal motivations accompanied work-related goals; "I like the conversations that come out of blogging: it's challenging."

Where to Blog?

We expected to find that the main decision when starting a weblog would be whether to blog internally or externally. However, more fine-grained choices and a broad variety of guiding criteria emerged, usually influenced by the goals for blogging, such as:

- *Access and visibility.* Who should be able to access the content? How easy will it be to find? Internal weblogs are good for sharing non-public information but have less exposure than an external weblog. Weblogs on official Microsoft servers are easily found by someone seeking Microsoft news; blogs on other external servers can be lost amid the many millions of other bloggers;
- *Affiliation with the company.* The choice of server can be influenced by a desire to have or avoid an explicit company affiliation. For some, their connection to Microsoft is a matter of credibility or pride; for others it adversely affects their image, leading them to be judged as Microsoft employees rather than for their expertise;
- *Freedom and control over technology or content.* Company-supported servers are an easy way to start blogging, but a self-hosted server (internal or external) can provide flexibility in configuring a weblog to fit one's preferences. Self-hosted or third-party platforms also raise fewer questions over the nature or ownership of the content.

What (Not) to Blog About?

With no formal policy, the lack of explicit rules creates a risk: each blogger is ultimately responsible for "being smart." Most weblogs we examined contain a disclaimer indicating that the content reflected the personal views of the author and should not be attributed to the company. But when an author openly associates with the company, the fine line between the personal and the corporate is blurred. Even weblogs primarily or exclusively focused on work are likely to have a personal touch, presenting information in an informal style and from an individual perspective. Many employees add personal comments to work-related notes or publish entries about hobbies, events in their private lives or opinions on non-work matters—suggesting that their readers "come to read the person, not the blog." Attitudes differ toward the propriety or desirability of mixing personal and work content. Some bloggers have two weblogs, one for work and one for personal content. Others share no private information online. Others see no problem with mixing work and private issues in a weblog that identifies their affiliation and often stress the role of personal information in providing context for work-related posts. Many struggle to identify what can be blogged about work, finding a grey area between the clearly confidential and the clearly publishable. In one group, bloggers praised clear communication from their management that identified "three topics you are not supposed to blog about." This provided clear boundaries while not curtailing the freedom to blog. For most it takes time, trial-and-error experimentation and reflection on internal and external feedback, to find what is comfortable for blogger, readers, and the company, trying to balance conflicting interests; one blogger said, "I fight with myself as a writer on behalf of Microsoft." Some respondents started conservatively and grew less so over time. Many described specific incidents that showed where to set boundaries. One mentioned intentionally writing a series of provocative posts to test the limits. Bloggers were challenged about posts by others, including people at higher levels. The relationship with the immediate manager was often identified as critical, in getting a blessing to start a weblog, negotiating acceptable uses, or seeking support in cases of unexpected negative effects of a post.

Blogging as Part of a Job

Given the time demands and work-related implications, how was blogging integrated into the day job for which a person was responsible? For a few, blogging eventually became an official part of their job. Indeed, in one case 15 hours per week were formally devoted to blogging. However, in most instances it is less dramatic. Some bloggers justified spending some work hours reading or writing weblogs by showing the impact on other responsibilities. Others did not make blogging a formal objective but raised it during annual performance appraisals as an extra work-related activity: "It's not explicitly part of my objectives, it's a means to an end," said one. A few bloggers strove for a complete separation of job responsibilities and blogging, even for primarily work-related blogs, to maximize their flexibility and freedom in posting.

Content Ownership

Despite the disclaimers, staff blogging about work, especially those using official servers, conceded that the company ultimately owned the content. This is consistent with the contracts governing the company's intellectual property rights, usually interpreted as applying to hardware, software and branding, but technically covering writing as well. However, not everyone agreed that all weblog content should be company property, but no one recounted a case where an ownership dispute had arisen, although their expressions of concern revealed uncertainty about the matter. For many, blogging involves personal initiative, investment and time, and could have long-term value in creating and maintaining an online reputation or as a record of thoughts and experiences. This played a role in the discussions about content ownership. Many would concede the right and need for the company to have access to the content of blogs closely related to specific products yet want to ensure their own access should they leave the company: "If they said they would delete it, I'd be thinking why am I blogging here [on company server] and not externally?" (cf. p. 216 here) Some took the extreme position of wanting sole ownership of their words and hosted their blogs externally, blogged on their own time, or both.

Conclusions

Be cautious in drawing conclusions from a study of a single company. The weblog community we observed is young and the environment is a strong shaping influence. In addition, with the technology and its adoption at an early point, new products will change the infrastructure. Features of weblogs are being integrated into diverse applications. Nevertheless, this study identifies issues that can guide organizations in making effective use of social software.

For an employee, a weblog can provide a space to share passion for work, to document and organize ideas and work practices, to find and engage others inside and outside the organization. For an employer, this can result in accelerated information flow, increased productivity, enhanced reputation and customer engagement but also in greater dependence on personalities, less control over the corporate face to the outside world, and possible challenges to hierarchy. A weblog, often started by personal initiative and supported by personal investment, can become an organizational asset, raising expectations and introducing risks. These considerations may motivate engagement with blogging, perhaps by providing support to maximize positive effects or by setting boundaries to minimize risks. Still, for many employees, blogging feels outside the corporate sphere of influence, even when clearly work-related. As a result, it is an arena for negotiation and interplay between personal and corporate interests.

We found disagreement as to what kinds of blogging made sense, though, and what kinds of content were appropriate. At our study site, key players in legal, public relations, and management were initially more negative than they were after more experience with the medium. If pushed to specify limits up front, an organization could be too restrictive and lose potential benefits. At the same time, bloggers should constantly consider limits and consequences—personal judgment and responsibility are inescapable elements of the activity. Employers and employees who take up blogging should anticipate that their practices will evolve. Their responsibilities may shift. Team relationships are affected. Experience and feedback change a blog; relatively formal blogs add personal touches, relatively personal blogs can take on project responsibilities. Issues arise when bloggers change jobs.

We expected perceptions and experiences around weblogs in Microsoft to be more confused than they were. We were surprised by the evidence of rapid evolution and growing sophistication. Perspectives had not converged; indeed, rapid changes in blogging practices raised new issues. A wide range of independent experiments was underway, accompanied by reflection and a keen sense of what was at stake. Blogging is about observing, reflecting, and commenting on surrounding activities, so perhaps this should not have been surprising.

Where encouraged, employee weblogs will change how work is organized and how authority is distributed by fostering direct communication across organizational boundaries, from employee to customer, and across group boundaries within organizations. The policy of "be smart" is telling; it becomes more important to have employees who are broadly informed. As we learn to exploit powerful new digital technologies, we may see significant changes in organizational forms; weblogs may be a manifestation of such change. Indeed, passion-driven, decentralized, and bridging personal and work contexts, employee blogging represents only one of the Web 2.0 technologies currently entering enterprises after adoption by consumers. Although our study does not provide all the answers, it indicates changes in workplace literacies that those technologies are likely to bring.

- Personal passions have a legitimate place at work. Personal stories and voices yield trusted relations. People are more likely to believe another human being than an organization or a computer. Showing emotions, telling personal stories, being passionate in hierarchical environments could be a challenge, but it is becoming an essential part of work.
- Transparency is here to stay. Weblogs provide a visible, often public, trace of one's expertise, actions and mistakes: what is written may stay "out there" forever and be searched, aggregated, transformed and

linked back to the author. When there is no way to escape one's past, it is essential to learn how to make mistakes in public and how to handle them gracefully.

- Visibility can turn into information overload. Being visible as a weblog author might extend one's reach but may also bring an unexpected explosion in communication as a result. With its low threshold for online publishing, blogging brings into public spaces ideas and stories previously hidden in private collections. Some of them are relevant and reliable, but most are extraneous, incomplete and not interesting, so important signals might be easily lost in the increased "noise." Complaints of information overload through blogging are the symptoms of navigating in the world of information abundance with habits and strategies learnt at times of information scarcity.

- Everyday routines matter. Unless one has nothing else to do, blogging survives only if integrated into the everyday world. Starting a blog is easy, continuing requires more—embedding the activity into one's information routines, work processes and interpersonal practices.

- Microactions aggregate. Blogging is about microcontent—publishing small pieces of thought and commentary, anchored with permalinks and carried away by feeds. However, the real value is not at the post level—ecosystems between blog posts and connections between their authors are more interesting and more important. Counting and measuring visible traces are tempting, but knowledge, reputation, and relations can escape rankings.

- Authority becomes fluid. Formal hierarchies of one kind or another do remain, but blogging provides alternative routes of influence.

In the end, our study shows that when it comes to blogging within/about their work contexts, individuals make judgments, take risks, and take responsibility for what they blog.

References

Adar, E., & Adamic, L.A. Tracking information epidemics in blogspace. *Proc. Web Intelligence 2005*.

Baskerville, R., & Dulipovici, A. (2006). The ethics of knowledge transfers and conversions: Property or privacy rights? *Proc. HICSS-39*.

Bishop, T. (2003, October 30). Microsoft fires worker over weblog. *Seattle Post-Intelligencer*.

Brown, J.S. (1983, December, 15). When user hits machine, or When is artificial ignorance better than artificial intelligence? *CHI'83 plenary address*, Boston, Massachusetts.

Cone, E. (2005, April 5). Rise of the blog. *CIO Insight*. April. Retrieved April 15, 2008 from: http://www.cioinsight.com/c/a/Past-News/Rise-of-the-Blog/

Dudley, B. (2004, June 25). Bill Gates could join ranks of bloggers. *Seattle Times*. June 25. Retrieved April 15, 2008 from: http://seattletimes.nwsource.com/html/businesstechnology/2001964841_gatesblog25.html

Efimova, L. (2004). Discovering the iceberg of knowledge work: A Weblog case. *Proc. European Conference on Organizational Knowledge, Learning and Capabilities*.

Efimova, L. & Grudin, J. (2007). Crossing boundaries: A case study of employee blogging. *Proc. HICSS-41*.

Grudin, J. (2005). Communication and collaboration support in an age of information scarcity. In K. Okada, T. Hoshi, & T. Inoue (Eds.), *Communication and Collaboration Support Systems* pp. 13–23). Tokyo, Japan: OSI Press/Ohmsha Ltd.

———. (2006). Enterprise Knowledge Management and Emerging Technologies. *Proc. HICSS-39*.

Grudin, J., Tallarico, S., & Counts, S. (2005). As technophobia disappears: Implications for design. *Proc. Group 2005*, 256–259.

Herring, S.C., Scheidt, L.A., Bonus, S., & Wright, E. (2004). Bridging the gap: A genre analysis of weblogs. *Proc. HICSS 37*.

Jüch, C., & Stobbe, A. (2005, August 22). Blogs: The new magic formula for corporate communications? *Deutsche Bank Research, 53*, 22.

Kelleher, T., & Miller, B. M. (2006). Organizational blogs and the human voice: Relational strategies and relational outcomes. *Journal of Computer-Mediated Communication, 11*(2) article 1.

Lovejoy, T., & Grudin, J. (2003). Messaging and formality: Will IM follow in the footsteps of Email? *Proc. INTERACT 2003*, 817–820.

Lovejoy, T., & Steele, N. (2004). Engaging our audience through photo stories. *Visual Anthropology Review, 20*, (1), 70–81.

Nardi, B., Schiano, D., & Gumbrecht, M. (2004). Blogging as social activity, or, would you let 900 million people read your diary? *Proc. CSCW 2004*, 222–231.

Perin, C. (1991). Electronic social fields in bureaucracies. *Comm. ACM, 34*, 12(1991), 74–82.

Pickering, J.M., & King, J.L. (1992). Hardwiring weak ties: Individual and institutional issues in computer mediated communication, *Proc. CSCW 92*, 356–361.

Snell, J. (2005, May 16). Blogging@IBM. Monday, May 16. Retrieved April 15, 2008, from: http://www.ibm.com/developerworks/blogs/page/jasnell?entry=blogging_ibm

Udell, J. (2005, May 24). The Weblog as a project management tool. *Tangled in the Threads*, 2001.3.

Viégas, F. B. (2005). Bloggers' expectations of privacy and accountability: An initial survey. *Journal of Computer-Mediated Communication, 10* (3), article 12.

Zerfaß, A. (2005, January 27). Corporate Blogs: Einsatzmöglichkeiten und Herausforderungen. Retrieved April 15, 2008, from: http://www.zerfass.de/corporateBlogs-AZ-27015.pdf

CHAPTER TEN

Pay and Display

The Digital Literacies of Online Shoppers

JULIA DAVIES

'Everything was stories and stories was everything'[1]

I BOUGHT THIS CAR A WHILE AGO OFF EBAY. IT'S THE FIRST YANK IVE EVER HAD TO BE HONEST IM STRUGGLING TO COPE WITH A LEFT HAND DRIVE THE SIZE OF IT DOES NOT HELP. I KEEP FINDING MYSELF STRAGGLING THE OPPOSITE LANE AND IVE HAD SOME NEAR MISSES MUCH TO THE DISPLEASURE OF THE MISSIS. I WOULD NOT MIND BUT IM A LORRY DRIVER. I LOVE THIS CAR TO BITS AND IF IT WAS A RDH I WOULD NEVER SELL IT. I USE IT AS A SECOND CAR BECOUSE IM AWAY ALL WEEK I DON'T WANT TO LEAVE MY ML MERC IN A CAR PARK ALL WEEK. I ONLY DO ABOUT 2 MILES A WEEK IN IT. WELL TO BE HONEST I WONT BE SAD IF IT DOES NOT SELL.

This is the first part of a narrative that forms a "listing" or an advertisement to sell a "1984 Chevrolet" on eBay.co.uk It continues in the same vein, giving details of how the seller came to own the car:

I BOUGHT THE CAR FROM A LOVELY GUY FROM THE NORTH EAST (HI WATSON) HE USED IT TO PULL HIS CARAVAN HE HAD IT FOR MANY YEARS AND SPENT A FORTUNE ON IT. IT ORIGIONALY HAD A 5.7 DIESEL IN IT WHEN HE BOUGHT WHICH WAS KNACKERD BUT HE STILL GOT 2 YEARS OUT OF IT TILL HE FOUND AN ENGINE FOR IT. HE WAS TOLD BY EXPERTS THAT THE 6.2 CHEVEY ENGINE WOULD NOT FIT BUT AN ENGINER FRIEND OF HIS SAID HE WOULD DO IT. AND BYE GOD HE DID A WONDERFULL JOB IT FITS LIKE A GLOVE.

The characterization in the story is strong; it depicts affable men ("lovely guy"; "Hi Watson;" "enginer friend") who love cars ("I love the car to bits"); who spend time and money on their passion, and whose sense of identity is closely allied to the vehicles they drive. "Experts" are mentioned as is "an engineer" and so one feels that these are men to be trusted. As a woman reading this, I feel the seller is talking to other men, with inferences of sexual prowess, especially in a later phrase "it will pull anything except women." There is a sense of humor here, good-natured banter and the car itself also has a character; it has a biography, people who love it and a determination to survive! The rather quaint colloquialisms, "bye God," "fits like a glove" and the use of capitalization throughout give it a kind of old-fashioned, (non-scary) rough edge so that one feels one is buying from a "regular guy" rather than a slick-suited dealer who doesn't care about the car. The idea is that one nice guy will sell to another nice guy. At one point the seller seems almost (but not quite) unable to control his enthusiasm when he comments towards the end that "THIS CAR WILL LAST FOR YEARS ITS BUILT LIKE A BRICK S**T HOUSE." This is a skilled teller of tales and one who has a strong sense of audience and context. The story is engaging in its use of superlatives and straightforward style; it continues to double the length of what I show here, and this also lends an effusiveness and passion to the tale. Twelve photographs accompany the text; the car shown parked on a grass verge, low angled shots emphasize the dimensions of the vehicle, and closer shots focus on the engine and the boot. Potential buyers are told, "IF YOU HAVE ANY QUESTIONS I WILL GET BACK TO YOU ASAP AS I WORK AWAY ALL WEEK AND NEVER KNOW WHEN IM HOME" in a comment which gives further credibility, since it verifies that the seller is a lorry driver.

This is a powerful text (with the car selling at £300.00 over the asking price), with its consistent tone, its apparent honesty, its direct approach. Perhaps some readers may have focused instead on the spelling errors, capitalization and seemingly unorthodox approach. Yet literacy is about more than accuracy; literacy, I argue in this chapter, is a social practice. This particular narrative embraces the values of "community" as defined by eBay making this text effective within the eBay context, and these two strands will remain a key feature of this chapter. It explores how the narratives of eBay contribute to the sense of community on eBay and demonstrates how an important literacy skill for effective use of eBay is the ability to present oneself as a community member.

"Human beings are storying creatures" argue Sikes and Gale (2006, no page). Giddens (1991) points out that this storying impulse is implicated with our sense of who we are, how we "fit" with the rest of the world and with notions of identity. Narrative helps us to make connections between events, feelings and experiences, and we can express these narratives in many different ways using a whole range of resources at our disposal—be they written or spoken words, still or moving images. Digital technologies afford us a whole range of opportunities to present such narratives as well as to collaborate over stories, to impact and question each others' stories and to carry these over from online spaces into other areas of our lives.

In this chapter I want to show the connections between the storying "impulse," what Hardy calls a "primary act of mind" (1975, p. 4) and some of the activities I have seen embedded in the online trading site of eBay (www.eBay.co.uk and www.eBay.com). I consider how eBay provides an opportunity for people to become involved in narrative. I reflect on how multimodal narratives can be traced through the site; narratives of identity, of objects and of lives. I further argue that sites like eBay, where people gain glimpses, albeit edited, of others' lives, often engage us because of their apparent "everydayness" because readers can implicate themselves into the texts or write themselves into the scripts, becoming involved in identity performances to varying degrees and in diverse ways.

The Research Process

The data presented in this paper reflect aspects of my own participation; I have been a member of the site for about four years and in this sense could be described as an "insider." I have gathered data by looking at what is happening, tracing the processes that people engage in during the course of buying selling, exchanging and trading. This process of tracing involves an ethnographic

perspective (Bloome & Green, 1996). In addition I have used two group interviews with eBayers, talking to them about their practices. It has not been
possible to provide images from eBay as the site prohibits such practices; I
have also not approached people to interview online for the same reason. I have
been in email contact with an eBayer through her practices on another site
and have been able to use an image from her photostream. Authenticity, trust,
reputation and the notion of community are the espoused values of eBay, and
in this chapter I explore how these are enacted by eBayers. In carrying out this
exploration, the importance of narrative within the site has emerged.

Of course eBay is fundamentally about capital exchange; people buy and
sell, make profits and losses. Yet the interactivity *is* more than the transaction
of goods; there are social practices within, beyond and around the trading. As I
show, eBay presents itself as an online "community" space, encouraging friend-
ly interaction as well as trading. The social networking aspect of the site lends
itself well to the production of capital benefit, and many sellers invest time
in creating highly engaging texts which promote active textual engagement.
Some so-called eBayers and some non-participants have reported to me that
they dislike all the "community" aspects of the site, seeing it as "intrusive" and
"disingenuous" while others are drawn to the site to engage with others. Whatever the orientation of users to the social context of eBay, the social aspect is at
least to some degree unavoidable—and for others is a key feature. I see much
of this interactivity as engagement in what might be seen as the site's narrative,
where individuals win and lose; where there are "goodies" and "baddies"; where
people discuss the "lives of objects," their previous owners and witness different ways of living. The site could be seen as an online glass cabinet, quixotic
and full of curios and items reflecting contemporary cultural interests, but it is
more than this, for it is not a museum which preserves and holds things still; it
is a dynamic space. In this chapter then, issues of community, of identity and of
narrative are explored. This may seem to some, a long way from thinking about
"digital literacies"; however if one considers, as I do, that literacy is a social
practice, then all this is fundamental for understanding online literacy events,
and in this chapter, the interactivity of eBay as a case in point. I briefly explain
below, my position in relation to literacy.

New Literacy Studies and New Literacies

The production and consumption of texts are social; that is, texts are produced
out of particular contexts, and readers bring to those texts and contexts, mean-

ings derived from their own experiences, cultures and social position. That is to say, social aspects are irrefutably part of literacy practices, and meaning is not just mediated straightforwardly by textual codes (alphabetized representation, etc.) but are shaped by socio-cultural matters. Texts are socio-cultural constructs. In emphasizing these points, I situate this chapter within what has become known as the paradigm of the "New Literacy Studies," associated with theorists like Barton and Hamilton (1998), Cope and Kalantzis (2000) and Street (1993), who describe literacy as a social practice. "Literacy is primarily something people do; it is an activity located in the space between thought and text" argue Barton and Hamilton (1998, p. 3).

Being literate therefore involves an understanding not just of how to decode alphabetically, but also involves being aware of all kinds of social "stuff" that surrounds texts. One needs to decode cultural and social context clues as Lankshear and Knobel (2006) argue:

> From a sociocultural perspective it is impossible to separate out from text-mediated social practices the bits concerned with reading or writing (or any other sense of literacy) and to treat them independently of all the non-print bits, like values and gestures, context and meaning, action and objects, talk and interaction, tools and spaces. They are all non-subtractable parts of integrated wholes. "Literacy bits" do not exist apart from the social practices in which they are embedded and in which they are acquired. (p. 13)

As will be seen from the examples I offer from eBay, this kind of argument is difficult to dispute where readers and writers of texts are primarily trying to *do* something; they are involved in social acts—selling, bidding, presenting, explaining, persuading and so on. The work of the New London Group in conceptualizing "multiliteracies" is now well established (Cope & Kalantzis, 2000); multiliteracies take into account a full range of modalities as contributing to meaning-making so that visual, aural, and spatial patterns are accepted as being as meaningful as the linguistic mode. Shifts, indeed a broadening of the meaning of the word "literacy," have been partly due, since the proliferation of screen-based and other digital texts, to the escalation of the use of different modalities within single texts (Kress & Van Leeuwen, 1996, 2000; Van Leeuwen & Jewitt, 2001). In this chapter I draw on multimodal analytical techniques, informed by Kress and Van Leeuwen (1996, 2000), and Van Leeuwen and Jewitt (2001) treating eBay as a text in itself, discussing meanings inscribed within individual entries on the site, their relationship with each other and the site generally, as well as the way in which other functionalities of the eBay software such as the ratings system, blogs and discussion boards impact on meanings. Moreover, eBay, (like so many other websites), is not self-contained; it exists in relation to other online contexts and those who frequent the site often make reference to other online spaces, often connecting to them through hyperlinks. Thus other spaces become part of the whole eBay text; hypertext renders textual boundaries permeable, so texts leak into others, affecting the way in which they may be read. For example, I was looking at some photographs on a blog and found they linked directly to the blogger's eBay listings as well as to her photostream on a photosharing website. I began to see how the boundaries between the sites were blending, with each site "leaking" into the other (e.g., liebemarlene.blogspot.com; flickr.com/people/liebemarlene/; ebay.com/LIEBEMARLENE-VINTAGE).

Walker talks of "distributed narratives," of stories that are not self-contained. She explains that "they can't be experienced in a single session or in a single space. They're stories that cross over into our daily lives, becoming as ubiquitous as the network that fosters them" (Walker, 2004, p. 1). This chapter shows some of the ways in which boundaries are challenged, as narratives pass through on- and offline spaces, being carried through objects, words and images in complex ways. This process helps us to traverse spaces, engage with others and collaborate over text-making and meaning.

Further, this chapter deals with what it sees as instances of new literacies. That is to say, it looks at the kinds of literacies in which new digital technologies produce texts engaging people in new social practices. Typically, this means, as Lankshear and Knobel (2006) explain, that new literacies are produced in ways that are "more collaborative," "less individuated," "more distributed" and "participatory." New literacies are new partly because they involve new social things being done; they affect us as social beings. The affordances of online spaces, particularly those known as "Web 2.0" spaces enable individuals to produce narratives which can easily be extended to include others, to be amended, extended, elaborated and so on, in shared ways. One space can be syntactically connected to another through a hyperlink and texts can merge together or be contained within the other. New Literacies allow texts to be jointly authored and meanings challenged, extended, altered and collaborated, perhaps help promote affinities or even the notion of community. I have written elsewhere, for example, how shared meanings can be developed through image-sharing sites (Davies, 2006, 2007).

The concept of community is promoted by eBay, and this helps define "insiders" and by default, those who are "outsiders," especially those who reject or undermine behaviors honored by the community and articulated in eBay's terms and conditions or codes of conduct. A certain level of literateness is therefore required for success in eBay, and to miss some of the social cues may lead to social difficulties and even ostracism. "Literateness" in New Literacies then includes a need to understand how texts perform complex social acts that can be inclusive or exclusive.

So, What Is eBay?

Having set out some of the ideas which shape my position and perspective on eBay, I move now to look at the site more specifically. EBay is a virtual marketplace of awesome proportions—probably the most successful online shopping space to date. It is an international institution; a virtual shopping mall where goods are exchanged amongst disparately spaced traders, where some now make a living and where others look, but never buy or sell.

It is a space that many know about, even if they have never visited, and it is a space through and about which stories are told. eBay describes itself as:

> The world's first biggest and best person-to-person online trading community. It's your place to find the stuff you want, to sell the stuff you have and to make friends while you are at it. (eBay, 2007, no page)

It is a huge operation, having separate but linked sites in some 27 countries (eBay, 2007) and being accessible in many languages. eBay hosts innumerable traders and buyers, and it is not possible for its owners to know about all items that are bought and sold. Goods do not pass through a central store or audit. It is a multiply-layered site—with chat rooms, discussion forums, shops, advice spaces and a huge emporium of items up for sale in categorized listings. Listings usually comprise words and images and sometimes links to associated online shops. Most items are up for auction, while some can be bought more quickly at set rates under a "buy now" option. Contracts of exchange are negotiated between buyers and sellers while eBay itself provides the software and regulations through which trading occurs. eBay is a concept and a text which others help define and substantiate through their interactivity; it is a space that is both constituted of and produced by text.

While eBay was initially set up as an auction site, increasing numbers of eBayers have virtual shops and/or offer fixed prices for many goods. Initially mainly dealing in second-hand articles, eBay sells increasing numbers of new goods and the predominance of bric-a-brac and rare items is now less obvious although these goods maintain a very strong presence.

To sell or buy, one must first register and therefore have an email address; so-called eBayers must give themselves a username and agree to the site's terms and conditions. After completion of a transaction, that is, after the sale of goods and goods have been received, the software prompts eBayers to leave a "rating" for each other, reflecting the quality of the trading procedure. EBayers rate each transaction as positive, negative or neutral and can give a comment. For example, one of my transactions attracted this very effusive response: "A1 EBAYER-MEGA QUICK PAYMENT-A CREDIT TO EBAY A1+++++." This message gives a clear signal to others with whom I may interact in the future—that I am trustworthy. A whole array of similar comments would confirm this further. In this way eBayers accrue a reputation which others can see online and use to judge whether they are good to deal with or not. As it is beneficial to present a good "pedigree," eBayers tend to co-operate with this system, often going to great lengths to keep a good record "clean." For example it is usual to receive with one's goods an exhortation like the one I received inside a box of coffee cups "I will now go and leave you positive feedback and hope you will do the same for me :>) Have a great day." In this way, eBayers take part in a ritual which endorses the values of the site and which presents trading as a co-operative exchange of social and financial benefit. I will return to this topic in relation to a discussion of the broader notion of "community" within eBay.

In addition to this ratings system, eBayers can see at a glance each others' levels of experience, because the software automatically counts and reveals each eBayer's accumulated number of transactions through color-coded stars. Any registered user, or "eBayer," can look at all the ratings and comments left for any other eBayer, and it is upon this system that trust depends. In this way, eBay provides an inescapable biographical account of all eBayers; it is an audit trading and about reputation—and strongly signals what is valued in the "community." Thus eBayers collaborate with each other and with the site, to present a particular story using a value system (or Discourse) that is an intrinsic part of the site's design, but with which traders are complicit.

Participation Requirements

In terms of baseline skills needed to participate, I estimate the minimum to be the ability to:

- set up an email account
- decide on a user name
- register on the site (with the above)
- set up a Paypal account (or similar to pay and receive payment securely)
- understand how an auction works
- place a bid
- understand a contract is binding
- understand that some items can be purchased outside the auction on a "buy now" basis—at a set price
- understand the terms and conditions of ebay for selling, buying and using the site
- leave feedback

In addition sellers need to be able to:

- arrange goods to photograph—in appropriate spaces, clear lighting, in appropriate angles and careful focus
- make a digital image and upload it to eBay
- describe an item accurately, concisely and enticingly
- categorize the item using the taxonomy provided by eBay
- calculate costs of postage and packaging
- display postage and packaging costs unambiguously
- send and receive emails from customers
- answer questions from buyers about goods
- respond promptly when items are sold

In addition buyers need to know how to:

- locate an item
- understand the verbal description
- decipher any linguistic or graphic conventions (e.g., BNNW—Brand New Not Worn; BNWL—Brand New With Labels)
- read critically—e.g., "shabby chic" may mean "old and battered" to someone else; "Royal Doulton style" means it is *not* "Royal Doulton" but is similar in some way
- interpret images—understand lighting and angle may mislead
- email questions if unsure
- understand postage and packaging options

- make bids or choose to "Buy Now"
- monitor bids to check progress

Many eBayers have multiple additional skills, as with the listing for the Chevrolet, with specialist knowledge about the item and potential buyers, appropriate language and a way of attracting attention. Considering the skills needed it is perhaps a surprise so many people participate. Literally millions of items are exchanged hourly, from shoes, toys, clothing and curios; from the brand new to the very old; and from the cheap to the frighteningly overpriced.

Objects and Desires

Numbers of participants are increasing daily and competition is immense, not just amongst sellers, but also amongst buyers—who vie with each other to purchase goods in auctions at the best possible price. As items purchased are frequently secondhand, obscure goods are highly valued; items with interesting histories are prized, even where objects have unclear purposes and materiality. So successful has been the selling power of some vendors that eBay is often the topic of newspapers, such as when a student sold a single cornflake for £1.20, or an artist sold his soul for less than £12.00 (BBC, 2002). Clearly it is not the intrinsic value of such items that attracts buyers; it is taking part in an unfolding dramatic narrative often witnessed beyond eBay by a global audience as it is further narrated by newspapers, television and bloggers, etc. In these instances the symbolic value of goods and involvement in narrative is what is being purchased.

Drawing on interviews with individuals and focus groups, as well as from data on eBayers' blogs, Ellis and Haywood (2006) found that those who were "early adopters" placed high value on second-hand goods and their histories, on specialist knowledge of those goods. They also found that value was given to usernames that were both appropriate and imaginative. Looking at their evidence in terms of literacy, narratives and identity issues, it seems that high value is attributed to the reputation of the seller, the online space that the seller may have "furnished" and on what seems to be the meanings within the objects themselves.

There is a sense that for many eBayers the "cultural biography of things" plays an important role for them and their practices online. One of the informants I interviewed told me that:

I love getting things which someone else has owned. I like getting stuff which seems to have a story behind it. I got a Good Housekeeping's recipe book from 1952. It had bits of food on some pages and even a handwritten shopping list inside. It was great. I imagined an old lady having owned it and I was carrying it on, kind of thing. (Toni, June 2007)

This short description, reported as part of a face-to-face group interview with eBayers, is a reflection of how material goods acquire semantic significance through their provenance. The new owner, a woman in her fifties, told me she had wanted to acquire this recipe book as it was like one her mother had used. The book represented an element of her childhood, as well as carrying traces of meanings from someone else's life. There was a sense of these stories coming together through the book itself. When I refer to "provenance," drawing on Rose who describes it as a "social mode of meaning" (Rose, 2001, p. 38), I refer to how an item accumulates layers of meanings through its history. The previous owners and uses the item has been put to become significant in its present state. In writing about the purchase of second-hand goods more generally, Gregson and Crewe (2003) refer to ". . . . the rituals involved in transforming the commodity into one's own result in high levels of attachment and the creation of new forms of meaning." It accrues meanings through its "biography"; Stewart (1993) talks of "objects surviving their original contexts. . . . as traces of the way of life that once surrounded them" (Stewart, 1993, p. 144). In describing souvenirs particularly, she says "Once the object is severed from its origin it is possible to generate a new series, to start again within a new context . . ." (ibid., p. 152). Similarly I am interested in the way in which images of objects online, accompanied by text, can generate interesting possibilities for meaning-making. Each image of an item carries traces of meanings from its original context but acquires additional nuances and associations from its online space. That is to say, simply by being on eBay, an item acquires additional history; by being sold through a particular seller and being associated with other goods that seller has, also adds to the provenance of an item. This is illustrated through the *Chevrolet* story, with a kind of life-history approach having been taken and where the characters of its previous owners seem to have imbued aspects of their character onto the car as it passed through their hands. Indeed in the process the car's materiality had changed over time, and changes that might be seen by some as making a wreck are presented as being symbolic of care and value. By purchasing this car, the new owner would become a part of the ongoing story.

The voices of individuals on eBay are easily discernible. I have also indicated that a combination of narratives run concurrently, for example through the feedback system which reveals eBayers' trading biographies. I now contextualize this within a brief discussion of the term "Discourse" as a prelude to exploring the notion of community more closely with a suggestion that the "Discourse" of "community" on eBay is used to regulate trading.

The d/Discourses of eBay

I now want to consider how eBay promotes a particular "way of being" specific to the eBay context. I discuss further how the site channels eBayers' behavior and how eBayers work within and beyond those parameters making them either insiders or outsiders to the "community." I draw on the notion of Discourse described by Gee in order to reflect my understandings. Gee distinguishes between "discourse" with a small "d" and "Discourse" with a large "D." This is a useful distinction with the latter concerning language and other "stuff" which can be conceived as expressing particular belief systems, values, and understandings. Discourses with a large "D," comprise social practices, mental entities and material realities. Gee talks of "[w]ays of behaving, interacting, valuing, thinking, believing, speaking . . . that are accepted as instantiations of particular roles . . . by specific groups of people . . . Discourses are ways of being 'people like us'. They are ways of 'being in the world'; they are 'forms of life' " (Gee, 1996, p. viii). They are thus always and everywhere social and products of social histories. He continues:

> socially accepted associations among ways of using language, of thinking, of valuing, acting, and interacting, in the "right" places and at the "right" times with the "right" objects (associations that can be used to identify oneself as a meaningful member of a socially meaningful group or "social network") I will refer to as "Discourses" with a capital "D". . . . "Big D" Discourses are always language *plus* "other stuff." (p. 17)

I see eBay as a specific discursive space, comprising multimodal discourses; being recognized as a real "eBayer" means following specific discursive practices that are part of the Discourse of eBay—this involves selling and buying in the space and following the rules, but further to these basic behaviors, there are other ways in which eBayers discursively situate themselves within the site. These are the ways that might be described as the culture, the socially accepted practices which may not be expressly articulated but which nevertheless are socially codified practices.

In this next section I show how eBay as an institution values and promotes the concept of "community" and how this Discourse is adopted by users who weave narratives around this concept. I show how the ratings system is promoted by eBay and invoked by its users as part of the "community" Discourse and how narratives evolve around those who adhere to this and those who do not.

The Discourse of Community

As mentioned earlier, eBay emphasizes the idea of "community" and refers to itself as follows:

> The world's first biggest and best person-to-person online trading community. It's your place to find the stuff you want, to sell the stuff you have and to make friends while you are at it. (eBay, 2007)

The repetition of "person" acknowledges individuals and suggests that trading will be built on friendship; eBay addresses eBayers directly as "you" and implies shared ownership: "your place." This notion of community is articulated more specifically in a set of specific values:

- We believe people are basically good.
- We believe everyone has something to contribute.
- We believe that an honest, open environment can bring out the best in people.
- We recognize and respect everyone as a unique individual.
- We encourage you to treat others the way you want to be treated.

eBay is firmly committed to these principles. And we believe that community members should also honour them—whether buying, selling, or chatting with eBay friends. (eBay, 2007, no page)

Even-handed fairness and openness are stressed; the words "good," "honest" and "respect" are the assumed attributes of eBayers and help define "the community." These values are referred to across the site by members as well as by the eBay administration. The norms of "goodness" and "playing a role" are assumed; this is a kind of flattery, a faith in the reader of this text. It is however also quite manipulative since the corollary is that by not taking on the community values, the individual is the inverse of these values—"not good," "disrespectful" etc.

While eBay's values are explicitly articulated, they are also enacted through the various modes and media made available to members. For example, the existence of discussion boards is testimony to the idea of showing respect for others and the need to listen. eBay gives trading advice to all eBayers, thus demonstrating an even-handed approach through its openness. It hosts spaces where veteran eBayers can advise "newbies" on a special discussion board and even has purely "social" boards, such as "The Nags Head" where topics totally unrelated to eBay can be discussed.

Despite the fact that the discussion boards are ostensibly the domain of eBayers, a space where they can talk outside the restrictions of the trading rituals, it is often in the discussion boards that members will invoke the Discourses of community values. For example, in answer to one "newbie" talking in the newbie's discussion board, another eBayer explains, "Sorry, but those are the rules you have to accept if you list on eBay." And on a different discussion when someone asked for help because she had been sent an item that she regarded as "fake," she is told by a veteran, "the listing states that the seller doesn't know if they're the real thing or not . . ." and that "You had three days to withdraw your bid! If you had clicked on Help at the top of any eBay page you would have found the answer! All you can do now is pay up, and maybe try to re-sell them." Here the newbie is seen as contravening the rules and so the "community" enforces the rules set by eBay. Sadly, it seems this eBayer's dilemma was caused by her inability to read between the lines. Lankshear and Knobel (aka netgrrrl (12) and chicoboy26 (32)) explain:

> eBay is not only a shaper within a new technologies arena, but it is also an educator in that it teaches people how they should act within this new cyberspace; how they should act in relation to each other. (netgrrrl (12) and chicoboy26 (32), 2002, p. 20)

The discussion boards may seem, because of the "architecture" of the site, to be somewhat marginal. The routes towards them are away from the central trading spaces and so perhaps seem peripheral to eBay's interactivity. Yet these boards are places where the notion of community is repeatedly rehearsed and reiterated in displays of commitment to the rules and regulations. Stories about other eBayers and trading practices are in profusion. After one bad experience from a fake buyer, one eBayer was helped by another. After accepting the advice she comments:

> Oh, thank you so much, I have never met a dishonest ebayer before—but I just had a bad feeling about this deal! I shall follow the links as you suggested and try to reclaim the fees. Do I have to wait for ebay's response or can I relist immediately? Thanks for your help—best wishes

In this posting the eBayer carefully defers to experience and uses a number of politeness markers "thank you so much"; "Thanks for your help" and "best wishes." In doing so she sets up an opposition between herself and the "dishonest" eBayer. Conversely she invests trust in her advisor of whom she immediately asks further questions. In this way she presents as an ideal eBayer who assumes that all eBayers are good—unless proved wrong.

The measures that some eBayers take in ensuring equitable trading are extensive and go beyond the site itself. For example in one of the discussions, somebody asked for help in deciding whether to trust a particular buyer:

> Okay, so I listed a Nintendo DS with 15 games with a buyitnow format with best offer function, I listed it at £279, he then made an offer which I accepted, I have sent him an invoice and have asked him to contact me or pay, but he has done neither. So I checked his feedback to see what he has bought or sold, it turns out all his bought items are private and the sellers that sold the items are all selling daft things like 1p eBooks and stuff, so I think his feedback is made up of 1p eBooks, I know I don't have to send the goods because he hasn't paid, but the listing fees and FVF come up to over £11, so how can I regain the fees if he does not pay?

In a display of his previous experience and knowledge of how things work, the seller displays his expertise as an eBayer, before asking a question. He uses the term "FVF" (Final Value Fee) and describes how he had used a channel setup by eBay (the feedback system) as a checking mechanism. He reveals how this eBayer may be undermining the system and requests advice on how to get back his fees from eBay. Replies provide web addresses giving the procedures to follow in re-claiming fees; the faith in the systems is demonstrated here by the way in which the questioner begins with an explanation about how he had followed all the rules and the replies all providing displays of knowledge relating to procedures. The next day the story continued with the initial complainant saying he had carried out detective work online—and beyond eBay:

> I checked his Contact details, it says he was registered in the UK. Also another thing, I checked his postcode and address through google and it turns out that his address is a business complex that has a large number of dissolved business under it and other small suspicious company names registered there too (I mean in the same building, on the same floor).

The story continues with a link to a website detailing a catalogue of scams committed by this particular eBayer. This is a story that crossed several spaces and that was collaboratively constructed. The discussion took place over three days and with participants based disparately across the UK. Their determination was demonstrated by their frequent contributions of information and suggestions, and it seems that the aim was to preserve the notion of community upon which their successful trading depends. The end result is perhaps a triumph of social networking and demonstrates the ways in which online identities can be traced across sites. In this case it was to the detriment of the rogue dealer, but other users invest in the permeability of online borders as a way of presenting an online self that is pervasive, consistent and trustworthy.

I mentioned above the way in which liebemarlene became known to me as an online presence through her blog. I was interested in the way she accompanied photographs with text that seemed part autobiographical and part promotional, such as this entry:

> This outfit is a little different from a lot of the ones I've been wearing lately, a little less frilly and girly. I've been inspired unintentionally by all the college students and football games going on, I think, though it looks like I'm a little stuck in the '70s, with an outfit that's part *Love Story* and part *Welcome Back Kotter*. (http://liebemarlene. blogspot.com/2007/11/fall-in-athens.html)

In linking to her photosharing site (Flickr.com) I found her photostream is dominated by images that look like stills from films, with posed shots such as the one below (see Figure 10.1).

Figure 10.1. liebemarlene's Flickrstream. (Source: http://www.flickr.com/ photos/liebemarlene/1798893503/in/set-72157594275740700/)

By using a blog, a flickrstream and her own "shop" on eBay, liebemarlene establishes an online presence that buyers can peruse and gain a sense of her authenticity by cross referencing through the spaces. She has created a kind of distributed narrative and provides the links to each space for others to follow. Her online identity displays a consistent passion for fashion; she comes across not as "just a trader" but as someone who has fused her genuine love of design with trading; in one of her flickr images she comments "just wearing these boots one last time before selling them." Through the images and text she is able to provide several narratives at once; firstly the narrative suggested by the image—a kind of film still—with the one above suggesting South Carolina in the 1970s; secondly a narrative about herself and her relationship with the clothes; thirdly a narrative about herself as a trustworthy online trader. In all of this she embraces the wider Discourse of the eBay "community," which is about authenticity, friendship and trading.

Whilst all sellers find their own way of presenting an image of themselves as honest and trustworthy—some simply declare it—such as in the frequently seen phrase "Buy with Confidence—See my Feedback!!" while others enact it in different ways. Some provide close-up images of items, especially those showing craftsman's stamps or even designer labels or packaging. Others have elaborate stories to tell and it is up to readers to be able to negotiate their way through this mass of textual clues. eBay provides opportunities for creativity, but it is clear that interactivity is framed within the Discourses it set up in the first place. There is a blurring of the boundaries between "community" as an eBay Discourse and the activities which eBayers are involved in; a measure of the success of eBay is not just in the numbers of those who trade on their site, but also in the numbers of those who actively promote its values. There seems very little evidence of cynicism or dissent amongst these many supporters, but it is possible these are so effectively silenced that I do not see them.

Conclusions

In a description of the "magical" properties of language, Gee (1996, p. 9) describes how "our worlds" are built through language and other cultural properties. He provides a useful taxonomy of the way "language in action" works in an "active building process" and suggests that through our use of language we simultaneously construct six elements of reality, as follows:

1. The meaning and value of aspects of the material world

2. Activities
3. Identities and relationships
4. Politics (the distribution of social goods)
5. Connections
6. Semiotics (Gee, 1996, pp. 9–10)

The way these six "areas of 'reality,'" as Gee terms them, are manifest on eBay, could be described as follows:

1. By registering in the online space known as "eBay" and entering discussion forums, clicking through to look at item descriptions or going into online shops, individuals become "eBayers." The space, framed by eBay logos and set out in a particular eBay way—with subdivisions, such as a homepage, personalized spaces ("my eBay"), shops, Community pages and so on—means that people act in certain ways according to where they are.

2. By engaging in acts such as buying, selling, discussing in forums, rating others in terms of eBay codes, or by taking photographs to upload to an online space, individuals are following the code of eBay and carrying out eBay activities. They become "eBayers."

3. By giving themselves usernames, presenting items through images and text, using particular types of language and codes, eBayers act as eBay sellers and their manner of engagement with others positions not only themselves but those with whom they interact.

4. There are particular ways of behaving that are highly valued on eBay and certain codes of behavior are deemed to be appropriate whilst others are not. Some of these are codified on the site by the site administrators, whilst others are ways of behaving that have been developed over time through custom and practice and which are part of the cultural politics of the space. Some people acquire a good reputation and others do not. These are clearly defined within the space.

5. Contributions to discussion boards or items that are exchanged by eBayers are all part of a whole ongoing process of actions that are linked together. Over time there develops a kind of social history so that there are nuances attached to particular topics, items and language which accrue as part of continued social interaction and connectivity.

6. Particular words or phrases are used within the space that may not be used elsewhere or words may have specialized meanings in the space.

For example acronyms like BNNW (Brand New Not Worn) or terms like "shabby chic."

It is not just the existence of the online space that constitutes eBay, its materiality (albeit virtual) comprises the six elements above, which are all part of an overarching eBay Discourse. In addition to this, the objects or commodities exchanged also form part of the online text and the online space. As far as the eBay space is concerned, the articles bought and sold exist as texts; their materiality is represented through words and images and value is given to those objects according to how they are represented; the value of the eBayer who sells them (the "situated identity" of the seller); and the meanings they have accrued in the online space and beyond. These artifacts are conjured through words and images before they are sold and after they are sold and may be referred to again in ratings given to the seller. They therefore have a traceable history with a narrative that is enacted online from pre-sale, point of sale and post-sale feedback. (Some goods of course follow a different story, such as reappearing for sale again if they are not sold first time around.)

Hillis (2006) cites a number of narratives used to sell particular items; for example the brand new Playstation that was asserted to be for sale "as punishment" to a son, and a similar story about a Nintendo console being sold as a punishment to two boys. The similarity of the stories implies this is a ruse, but either way, the narrative is used as part of the accompanying information about the object. There is an implicit assumption that buyers will be interested and engaged by such details, and indeed there are many items that have sold on eBay simply because of the narrative, or an idea invested in an item rather than the intrinsic value of the commodity.

According to Hillis and colleagues, it is "The world's largest online market, eBay is a virtual setting where capital, desire, and identity converge" (Hillis et al., 2006, p. 1). I have shown how, through online participation in content production and consumption practices, those who are involved become textually constituted and part of an ongoing eBay narrative or "Discourse" (Gee, 1996). I have traced the social practices that people engage in during the course of buying, selling, exchanging and trading and offer a multimodal account of the textualized behaviors within this huge online "shopping mall."

Authenticity, trust, and reputation are key to those who participate within eBay, and I have explored how these are enacted. The structure of the site ensures that involvement in trading prompts participation in practices associated with the valuing of reputation using the "rating system" which is an integral part of the site's design. This process closely aligns the nature of the seller with the quality of goods sold. Far from resisting this system, where reputations of sellers and buyers are an important part of consumption and exchange, this feature of the site is highly valued by sellers and buyers alike. The invocation of reputation is a common feature in the "pitch" sellers use to promote their goods and is foregrounded in other parts of the site as well. Moreover, many participants use the affordances provided by additional online spaces in order to self-present, thus investing in a wider range of modalities to enrich their presentations of self and goods. This use of multiple online spaces allows potential buyers to trace a kind of online biography of sellers, where through a process of triangulation, authenticity of goods and of their sellers seems more possible. In turn, this process seems to infuse the meanings of the goods, which helps make the textual identities of sellers work almost as a kind of "brand." It is clear that many sellers and buyers on eBay are not only becoming acquainted with practices promoted by the site and able to be creative within the parameters on offer, but many are innovating new practices.

Drawing on the notion that literacy is a social practice, I have argued that the literacy skills required in order to participate successfully as a buyer or seller are highly complex. eBayers' text production and consumption practices require extensive skills in order to decode verbal descriptions and to read the photographic images of objects. They need to conduct relationships as buyers and sellers, involved not just in particular isolated transactions, but with an understanding that individual exchanges contribute to an accumulated biography of themselves that can be referred to later. eBayers also need to read critically; where relevant they need to be able to explore beyond eBay itself; they need to understand "community" rules and be able to discern eBay's social and cultural values and demonstrate they understand them through what they say as well as in what they do. This is a space where traders form a self-monitoring community, with an awareness of "being watched" in a manner evocative of Foucault's panopticon (Foucault, 1977).

Finally, the real world counterpart to eBay, the shopping mall located in geographical space, attracts a good number of window shoppers who peruse the place without buying or selling. There are those who look at eBay without participating, who "lurk" in online spaces. Traditionally these invisible individuals are unaccounted for, but perhaps a later project needs to examine those flaneurs of the virtual shopping mall, who look at others without being detected and who do not interact.

Note

1. Harry Crews' opening line from the film *Searching for the One-Eyed Jesus*. DVD: Home Vision Entertainment.

References

Barton, D. & Hamilton, M. (1998). *Local literacies: Reading and writing in one community*. London: Routledge.

Bowlby, R. (1993). *Shopping with Freud*. London: Routledge.

BBC News online. (2002). Soul sold for less than £12. Retrieved August 2007, from http://news.bbc.co.uk/1/hi/england/2051061.stm

———— (2004) News report: "Single cornflake sold on internet." Retrieved August 2007, from http://news.bbc.co.uk./1/hi/england/coventry_warwickshire/4137877.stm

Bloome, D., & Green, J. (1996). Ethnography and Ethnographers of and in Education: A Situated Perspective. In: J. Flood, S. Heath, D. Lapp (Eds.), *A handbook for literacy educators: Research on teaching the communicative and visual arts* (pp. 1–12). New York: Macmillan.

Cope, B., & Kalantzis, M. (Eds.). (2000). *Multi-literacies; Literacy learning and the design of social futures*. London: Routledge.

Davies, J. (2006). Affinities and beyond!! Developing ways of seeing in online spaces. *e-learning-Special Issue: Digital Interfaces, 3(2),* 217–234.

————. (2007). Display, identity and the everyday: self-presentation through digital image sharing. *Discourse: Studies in the Cultural Politics of Education. 28(4).*

DeCerteau, M. (1984). *The practice of everyday Life*. London: University of California Press.

eBay. (2007). *Online auction website*. Retrieved November 25, 2007 from eBay.co.uk and eBay.com

Ellis, M. R., & Haywood, A. J. (2006). *Full Report of Research Activities and Results 'Virtually second-hand: Internet auction sites as spaces of knowledge performance'* Report to the ESRC for award no: RES-000-23–0433. Retrieved October 2007, http://www.essex.ac.uk/chimera/content/pubs/pubs/Virtually%20second-hand%20ESRC%20research%20report.pdf

Foucault, M. (1977). *Discipline and punish: The birth of the prison*. New York: Pantheon Books.

Gee, J. (1996). *Social linguistics and literacies: Ideology in discourses*. 2ⁿᵈ Ed. London: Falmer.

Giddens, A. (1991). *Modernity and self identity: Self and society in the late modern age*. Oxford: Polity.

Godwin-Jones, R. (2006). Emerging technologies: Tag clouds in the blogosphere: Electronic literacy and social networking. *Language Learning & Technology, 10(2),* 8–15. Retrieved August 2007, from http://llt.msu.edu/vol10num2/emerging/

Gregson, N. & Crewe, L. (2003). *Second-hand cultures*. London: Berg Publishers.

Hardy, B. (1975). *Tellers and listeners: The narrative imagination*. London: The Athlone Press.

Hillis, K. (2006). A space for the trace: Memorable eBay and narrative effect. *Space and Culture. 9(2),* 140–156.

Hillis, K., Petit, M., & Scott Epley, N. (Eds.). (2006). *Everyday eBay: Culture, collecting and desire.*

London: Routledge.

Klaffke, P. (2003). *Spree: A cultural history of shopping.* Vancouver: Arsenal Pulp Press.

Kress, G., & Van Leeuwen, T. (1996). *Reading images: The grammar of visual design.* London: Routledge.

———. (2000). *Multimodal discourse: The modes and media of contemporary communication.* Oxford: Edward Arnold.

Kuhn, A. (1985). *Family secrets: Acts of memory and imagination.* London: Verso/Langelier.

Lankshear, C., & Knobel, M. (2006). *New literacies: Everyday practices and classroom learning.* 2nd edn. Maidenhead: Open University Press.

Miller, D. (1998). *A theory of shopping.* London: Routledge.

Miller, D., Jackson, P., Thrift, N., Holbrook, B. & Rowlands, M. (Eds.). (1998). *Shopping, place and edentity.* London: Routledge.

netgrrl★ (12) and chicoboy26★ (32) (2002). What am I bid? Reading writing and ratings at eBay.com. In Snyder, I. (Ed.), *Silicon Literacies: Communication, Innovation and Education in the Electronic Age.* London: Routledge.

Rose, G. (2001). *Visual methodologies.* London: Sage.

Sikes, P., & Gale, K. (2006). *Narrative Approaches to Education Research.* From, online papers on *Research in Education* for The University of Plymouth, UK. Retrieved October 25, 2007, from http://www.edu.plymouth.ac.uk/resined/narrative/narrativehome.htm#Narrative%20 accounts%20of%20lives

Stewart, S. (1993). *On longing: Narratives of the miniature, the gigantic, the souvenir, the collection.* Durham, NC: Duke University Press.

Street, B. (ed.) (1993). *Cross-cultural approaches to literacy.* Cambridge, UK: Cambridge University Press.

Trodd, Z. (2006). Reading eBay: Hidden stories, subjective stories, and a people's history of the archive. In K. Hillis, M. Petit, & N. Scott Epley (Eds.), *Everyday eBay: Culture, collecting and desire* (pp 77–89). London: Routledge.

Van Leeuwen, T., & Jewitt, C. (2001). *Handbook of visual analysis.* London: Sage.

Walker, J. (2004). Distributed narrative. Blog post at http://jilltxt.net/txt/distributednarrative. html. Retrieved April 22, 2008.

Zukin, S. (2004). *Point of purchase: How shopping changed American culture.* Routledge: London.

CHAPTER ELEVEN

Digital Literacy and Participation in Online Social Networking Spaces

MICHELE KNOBEL AND COLIN LANKSHEAR

Introduction

Purpose-specific social networking sites are an early runaway success story among Web 2.0 social software applications. Their rapid uptake around the world and the diverse and complex features associated with participating in social networking spaces mean discussion of digital literacy would be incomplete without addressing this dimension of everyday literacy practice. Here, we discuss participation in social networking spaces from the standpoint of a particular conception of literacies. This defines literacies as "socially recognized ways of generating, communicating and negotiating meaningful content as members of Discourses through the medium of encoded texts."

Social networking sites and services have been studied from several theoretical perspectives to date. These include network theory (e.g., Paolillo & Wright, 2005), signaling theory (e.g., Lampe, Ellison, & Steinfeld, 2007), human geography theory (e.g., Humphreys, 2007), social contract theory (e.g., Snyder, Carpenter, & Slauson, 2006), and the sociology of groups (e.g., Baym,

2007; Boyd, 2008). Danah Boyd and Nicole Ellison claim that the "bulk of SNS [social network site] research has focused on impression management and friendship performance, networks and network structure, online/offline connections, and privacy issues" (2007, p. 11). Boyd and Ellison further identify an emerging range of new research foci within the study of social networking spaces, including race and ethnicity, gender and sexuality, religion, and civic engagement.

Few studies to date have examined social networking spaces from a literacy/literacies perspective. Of the studies identified by Boyd and Ellison (2007), only Dan Perkel's (in press) examination of MySpace users copying HTML code from other people's profiles and pasting it into their own profile pages emerges as a clear example of research examining social network sites in terms of digital literacies. Perkel uses a socio-technical theory of literacy, developed from the work of Andrea diSessa (2001). He argues that educators need to pay attention to new representational forms—such as those practiced within MySpace profiles—in order to consider how new literacy practices and processes of "re-using a diverse array of media" may be signalling "a deep shift in how people engage with one another" (no page). This chapter seeks to augment Perkel's pioneering study by bringing a sociocultural lens to bear on the field. We believe that a sociocultural theory of literacy can illuminate understanding of participation in social networking sites by fleshing out the "literacy as social practice" dimensions of these spaces.

Social Networking Sites

The kind of online social networking spaces discussed here requires specialized interfaces that help participants manage information about themselves, facilitate connections with selected others through quick links to their profiles and automated updates, etc., and help them manage diverse interpersonal interactions with others (e.g., text, image, video and audio messaging systems; testimonial spaces; song clip sharing facilities; interactive games; quizzes; photo sharing and tagging). Popular examples of social networking sites include Facebook.com, MySpace.com, Friendster.com, BlackPlanet.com, MiGente.com, Cyworld.com, Bebo.com, Orkut.com, and Hi5.com, among many others. We are *not* concerned here with the diverse array of other online resources and facilities that can be used for social networking purposes or the practices associated with them—such as exchanging email or instant messages, or posting to discussion-boards, operating a blog, or participating in "affinity spaces" (Gee 2004, 2007).

Boyd and Ellison (2007, p. 10) identify the rise of social networking sites with "a shift in the organization of online communities." Within a burgeoning internet culture in which "websites dedicated to communities of interest still exist and prosper" the subset of sites that comprise social networking spaces "are primarily organized around people, not interests." (ibid.) Whereas public online communities that sprang up in the 1980s and 1990s like Usenet and public discussion forums "were structured by topics or according to topical hierarchies," social networking sites "are structured as personal (or 'egocentric') networks, with the individual at the center of their own community" (Boyd & Ellison, 2007, p. 10). This characteristic also distinguishes engagement in social networking sites from participation in affinity spaces, since the latter are organized primarily around shared endeavors, rather than around identity and relationships with individuals at the center of their own networks.

Social networking sites typically share three general defining characteristics. They

> allow individuals to (1) construct a public or semi-public profile within a bounded system, (2) articulate a list of other users with whom they share a connection, and (3) view and traverse their list of connections and those made by others within the system. (Boyd & Ellison, 2007, p. 2)

A range of informative accounts of social networking sites already exists. Those by Kumar, Novak and Tomkins (2006), Boyd and Ellison (2007), and Boyd (2008) provide helpful insights into the history and sociology of social network spaces and offer broad orientations to the development of social networking sites since SixDegrees.com and similar sites were launched in 2001 (see Boyd & Ellison, 2007). In the remainder of this section we will provide just enough preliminary concrete background detail to meet our immediate purposes for exploring participation in social networking sites as forms of engagement in practices of digital literacy. We use Facebook.com since this is the social networking site with which we are most familiar, and it exemplifies the defining characteristics of social networking sites identified above. This background provides a "static" description of the Facebook interface. After outlining our view of digital literacies we will take some typical "action shots" of participation in Facebook. These will provide data from which we seek to understand popular participation in social networking sites in terms of digital literacy practices.

Facebook.com: Interface and Functionality

Facebook (http://www.facebook.com) is one of the best-known examples of social networking spaces. Launched in 2004 by Mark Zuckerberg as a social networking site open only to Harvard University students, it was subsequently extended to college students elsewhere in the U.S. In 2005 it opened to people with email addresses from any university (e.g., having a domain such as "edu," "ac.uk," or "edu.au"). In 2006, Facebook expanded into the public domain. By mid-November 2007, it was credited with 54 million active members worldwide (Wikipedia.org, 2007). Ellison, Steinfeld and Lampe (2007, no page) cite data from May and Kwong (2007) indicating that in 2007 Facebook users were "generating 1.5 billion [Facebook] page views each day." Ellison and colleagues further report that the "site is tightly integrated into the daily media practices of its users: The typical user spends about 20 minutes a day on the site, and two-thirds of users log in at least once a day" (Cassidy, 2006; May & Kwong, 2007).

The following general description of a Facebook profile (the "touchstone" page for participating in and navigating Facebook) conveys the look of this particular social networking interface in November 2007. In this description "you" refers to the owner of a Facebook page.

Figure 11.1: A Clip from a Facebook.com Page.

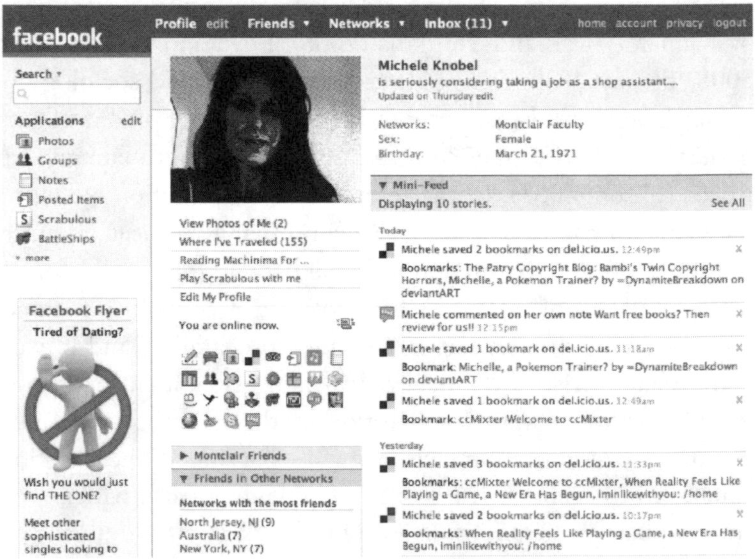

(Facebook © 2008)

Facebook's profile page is divided into three columns with a menu bar across the top (see Figure 11.1). The first column includes a search function for finding "friends" or groups already "on" Facebook and for inviting friends to join Facebook. It also includes a list of applications that are part of the profile page by default (e.g., for displaying photos, group memberships, notes), or that have been added by the "owner" of this page (e.g., links to games being played with other "Facebookers," to virtual libraries, to quizzes, or to travel maps). This first column also hosts commercial advertising.

The second column is headed up with a profile photo posted by the page's owner, followed by a set of hyperlinks to content areas contained within this profile page or to content management functions accessible only to the owner of the page (e.g., "edit my profile"). Beneath these hyperlinks a set of icons link to content and interactive applications within the Facebook universe. Below this again is a list of "networks" to which your friends belong. A network can be a university, a workplace, a city, or a country. This column can also include links to Facebook groups you have subscribed to (e.g., "People Who Always Have To Spell Their Names for Other People", "New Literacies and Social Practices"), and a box for displaying notes or announcements on your profile page.

The third—and largest—column includes personal information displayed according to your disclosure preferences (e.g., birth date, telephone numbers, email addresses). This personal information section includes a "status update" function where you key in text to finish the sentence: "Your Name _____" (e.g., Colin is drinking very strong coffee). This status update is easily changed and visible to everyone within your friendship network and can alert people immediately to your emotional status or to things happening in your life (where you are in the world, what you're currently doing, how some event is panning out).

This third column also includes by default what is called a "mini feed," which automatically tracks what you and others in your friendship network are doing within the Facebook universe (e.g., winning a game, high scoring on a quiz, writing a public message on someone's wall). This mini feed is organized chronologically from the most recent event to the least recent event and is shaped by the applications added to each profile page. For example, Michele has added a bookmarking function to her Facebook page that synchronizes her Del.icio.us bookmarks (see http://del.icio.us) with Facebook. Each bookmark she saves is also reported in her mini feed and friends can see and visit the online websites she bookmarks if they wish. She also has added an application that tracks her blogs and displays short summaries of recent blog posts to her profile page for others to read at a glance and follow up if interested (this application also posts what friends—and their friends—are blogging about). Each added application is contained within a "box" on her profile. Most boxes within the second and third columns of the profile page (e.g., the boxes for Scrabble, your Aquarium, your Virtual Bookshelf) can be moved around, reordered, hidden or deleted as the owner pleases. The remainder of the third column is filled with personal details (e.g., interests, favorite movies, favorite television shows) and your online life (e.g., space for including hyperlinks to your blogs, websites, photo display account, etc.). A popular feature of this third column is some version of a "wall" or comment space where friends can post public messages using text, video, audio, still images, etc.

The menu across the top of the profile page lets you access all your Facebook friends with just a few mouse clicks. "Friends" are established via mutual agreement to link to each other's profile page. Friends can be existing friends, acquaintances or colleagues found or acquired through the Facebook search function. Friends can also be strangers "met" via your friends' own friend lists, through group memberships, or through shared applications like the Virtual Bookshelf application or interactive games. This same menu bar shows which of your friends have recently updated their profile page and which are currently online. The menu bar provides access to functions that let you edit profile information, to lists of networks you (and your friends) belong to, and to your message center (where friends or groups can leave you private messages).

This "static" account of key features of the Facebook interface and the things it enables users to do conforms precisely to the three criteria identified by Boyd, Ellison and other researchers as characterizing social networking spaces. Facebook allows users to build a profile page within a distinctive online service or utility, to connect with others in mutually agreed-upon ways, and to view and "traverse this list of connections and those made by others within the system" (Boyd & Ellison, 2007, p. 2).

Facebook, like most other social networking spaces, is profile-driven. These profiles are by default hidden from general visitors to the site who are not part of that user's social network. Even members of Facebook itself can only see a profile photo, the person's name and the names of key networks with which each person is affiliated when browsing within the site (unlike MySpace, for example, where large parts of each profile are public beyond friendship networks). Friends—users formally connected to a person's profile—can see your full or partial profile, depending on your preferences. Facebook is bounded in the sense that you need to register with the service to participate. It is also bounded in the sense that while you are able to add Facebook-endorsed content and post messages or notes within your profile page space, you cannot add additional pages of your own making, or upload your own content beyond text, images and videos (unlike MySpace, which allows users to add wallpaper images, embed songs etc.). Facebook also articulates and manages your friends list. It helps you establish connections with people you want to add to your friends list, alerts you to changes on your friends' profile pages, enables you to view your friends' friend lists, and provides a range of services that let you traverse your social relationships within this space and interact socially with your friends (e.g., public wall posts, private messages, playing games together, taking comparison quizzes, showing what you and your friends are currently reading or watching, sending small digital icons to each other, etc).

Literacies

We briefly consider here some relevant aspects of the four key concepts in our definition of "literacies" as "socially recognized ways of generating, communicating and negotiating meaningful content as members of Discourses through the medium of encoded texts."

a. Socially Recognized Ways

The idea of "socially recognized ways" is close to the concept of "practice" developed by Scribner and Cole (1981) in their classic account of literacy as social practice. They defined practices as "socially developed and patterned ways of using technology and knowledge to accomplish tasks" (p. 236). A practice is a "recurrent, goal-directed sequence of activities using a particular technology and a particular system of knowledge" (p. 236). Whenever people participate in tasks that involve them in pursuing "socially recognized goals" and in making use of "a shared technology and knowledge system" to achieve these goals, they can be seen as "engaging in a social practice" (Scribner & Cole, 1981, p. 236). Applying knowledge in conjunction with some technology to accomplish tasks always involves "co-ordinated sets of actions," which Scribner and Cole refer to as "skills." Practices, then, comprise technology, knowledge and skills organized in *ways* that participants recognize, follow, and modify: they are organized (or co-ordinated) and deployed in *socially recognized ways*.

In applying this concept of practice to literacy, Scribner and Cole approach literacy as "a set of socially organized practices which make use of a symbol system and a technology for producing and disseminating it" (1981, p. 236). Literacy is not a matter simply of knowing how to read and write a particular kind of script but, rather, a matter of "applying this knowledge for specific purposes in specific contexts of use" (ibid.). Hence, literacy is really like a family of practices—literacies—that include such "socially evolved and patterned activities" as letter writing, keeping records and inventories, keeping a diary, writing memos, posting announcements, and so on. These all vary to some extent from one another in terms of the technologies used (pencil, typewriter, pen, font options, the kind of surface "written" on); the knowledge drawn upon (formatting conventions, use of register, information about the topic, audience), and their skill requirements (hand-eye co-ordination, using a mouse).

Here we explore social networking as such a member of a family of literacy practices. The symbol system in social networking is highly complex. It is not a unitary symbol system like alphabetic print writing. Online social networking employs a mix of symbol systems and modes rendered seamless by digital code and integrated hardware and software.

b. Meaningful Content

Generating and communicating meanings, inviting others to make meaning from our texts, and doing so with others in turn, can only be done by having something to make meaning *from*; namely, a kind of content carried as "potential" by a text and actualized as *meaningful* content through interaction with the text by its recipients. If there is no text there is no literacy.

Ideas of "meaningful content" can be wider or narrower, looser or tighter, depending on how close one stays to "literality" and to the idea that text is "self-contained." We take quite a loose approach, which puts much weight on the complexity and richness of the relationship between (new) literacies and "ways of being together in the world" (or, "Discourses," see Gee, 2004). When looking at somebody's weblog one might well find that much of the meaning one makes from the content has to do with who one thinks the blog writer *is*: what they are like, how they want to think of themselves, and how they want us to think of them. Likewise, a particular text that someone produces might well be best understood as an expression of wanting to feel "connected" or "related" right now rather than as a statement of literal information. The meaning we make of the content might hardly be literal at all. It might be almost entirely *relational*, in the sense of being about expressing solidarity or affinity with certain other people.

c. Encoded Texts

By "encoded texts" we mean texts that have been "frozen" or "captured" in ways that free them from their immediate context and origin of production, such that they are "(trans)portable" and exist independently of the presence of human beings as bearers of the text. The kinds of codes employed in literacy practices are varied and contingent. Literacies can involve *any* kind of codification system that "captures" material for generating, communicating and negotiating meaning in the sense we have mentioned. Literacy includes "letteracy" (recognition and manipulation of alphabetic symbols), but in our view goes far beyond this. In our view, someone who "freezes" language as a digitally encoded passage of speech and uploads it to the internet as a podcast is engaging in literacy. So, equally, is someone who photoshops an image; whether or not it includes a written text component.

d. Participation in Discourses

Discourse can be seen as the underlying principle of meaning and meaningfulness. We "do life" as individuals and as members of social and cultural groups—always as what Gee calls "situated selves"—in and through Discourses, which can be understood as meaningful co-ordinations of human and non-human elements. Besides people themselves, the human elements of co-ordinations include people's ways of thinking, acting, feeling, moving, dressing, speaking, gesturing, believing, and valuing. Non-human elements of co-ordinations include such things as tools, objects, institutions, networks, places, vehicles, machines, physical spaces, buildings. And "[w]ithin such co-ordinations we humans become recognizable to ourselves and to others and recognize ourselves, other people, and things as meaningful in distinctive ways" (Gee, 1997, p. xiv).

Literacies can be seen both as elements of co-ordinations, and as themselves co-ordinations that are parts of Discourses. Meaning-making draws on knowledge of Discourses; that is, on insider perspectives, and meaning-making thus often goes beyond what is "literally" in the sign. Part of the importance of defining literacies explicitly in relation to Discourses, then, is that it speaks to the meanings that insiders and outsiders to particular practices can and cannot make respectively. It reminds us that texts evoke interpretation on all kinds of levels that may only partially be "tappable" or "accessible" *linguistically*.

Digital Literacies

The complexity of literacy practices as myriad social phenomena invites useful forms of classification. From this perspective, *digital* literacies, quite simply, involve the use of digital technologies for encoding and accessing texts by which we generate, communicate and negotiate meanings in socially recognizable ways. In the case of online social networking these technologies are purpose-designed Web 2.0 internet technologies, comprising resources/utilities like those used by members of Facebook, MySpace, Friendster, etc., communities.

From this standpoint, a key purpose served by talk of digital literacies is to focus attention on the symbol system component of Scribner and Cole's account. As Bill Cope and Mary Kalantzis (2005, pp. 200–201) note in an interview with Colin Lankshear, digital technologies have reduced the basic particle of composition involved in communicating meaning via texts from the level of characters to a level *beneath* the character. Text rendered on the screen is reduced to pixels, with keystrokes building visual representations out of pixels; "You click for 'A' and you click for red" (ibid., p. 200). This involves a logic of representing characters. The basic unit is not the characters, but whatever programming language and group of pixels constitutes a particular character. Moreover, if we push back behind pixels "the same compositional stuff produces sounds as well" (ibid.). This has massive implications for human communication.

> [L]anguage, visuals and sound . . . are all being manufactured in the same raw material on the same plane, on the same platform. Give human beings the capacity to communicate in any way and they'll take it up. We are witnessing a huge turn away from the dominance of alphabetical language; a turn away from privileging isolated written language; and a turn towards the visual. This turn towards the visual can partly be understood in terms of the fact that in the current context of globalization, when languages are not mutually intelligible, you have to carry things visually. [Currently] a lot of text, like the instructional manual for a digital camera or the signs around an airport, involves the meaning being carried by icons. This is an attempt to reduce some aspects of language to visual schemas. (ibid., p. 201)

As we will see in the case of participating in Facebook, the technology of digital literacies affords a symbol system of unprecedented scope, sophistication, and complexity to those with the means at their disposal. This facilitates intriguing layers of communicative purposes that can be realized simultaneously, with an ease that encourages experimentation, creative innovation and playfulness, and in ways that make "bottom lines" out of what might well have been considered "luxuries" and self-absorbed excess barely a decade ago.

Understanding participation in social networking sites in terms of digital literacy practices involves considering some of the socially recognized ways in which people go about generating, communicating and negotiating meaningful content through the medium of digitally encoded texts of various kinds in contexts where they interact as members of Discourses. We draw on data derived from some typical examples of participation in action using the cases of Michele (co-author of this chapter) and Chris (a friend who works in the computer and music industries and who has a well-established life online).

Friendship in Action on Facebook: Michele and Chris as "Network Centers"

Two Networks and Their "Protagonists"

In the case of Facebook, "network" can be understood in at least three different ways. First, networks may comprise the formally identified networks afforded by the Facebook utility itself (traversing countries, regions, cities and universities). Second, they may comprise the social networks within one's list of friends (family, work mates, high school friends), which are not necessarily visible as formal networks. Third, networks may refer to membership of groups formalized as such within Facebook.

(a) Michele

Michele's formal networks containing most friends—deemed by Facebook to be the "strongest" networks—comprise Australia (birthplace), North Jersey (where she lives mostly), New York and Boston (where friends reside), Montclair State University (workplace), and McGill and Columbia Universities (where friends study or work). Those of her friends indicating network membership in their profiles belong collectively to over 60 different networks.

Michele's informal social networks—not so easily spotted by outsiders—include family members, friends with whom she socializes in person, school mates from high school days, university colleagues in the U.S., Australia, Finland, and England, doctoral students from diverse universities met via friends and conferences, ex-students from universities in the U.S. and Canada, and someone living in Mexico met via a virtual bookshelf Facebook application.

These networks are maintained largely through "super-poking," game-playing, wall posts, public notes, private messages, and comments on photos posted to Facebook albums. A number of people in her network she has never met face-to-face.

At time of writing, Michele belonged to 15 "groups" on Facebook. These are all driven by shared interests and established by Facebook users themselves. Michele's memberships reflect a range of her interests. Most tap her academic interest in literacies and/or digital technologies (e.g., "Language, Literacy and Power," "Language Learning with New Media and Games," the "Harvard Interactive Media Group," and the "MIT Videogame Theorists" group). Some groups were developed around conferences (e.g., "ROFLcon"—a conference on memes—and "New Literacies and in Social Practices"—a conference on media education). Other groups focus on social issues (e.g., the "09–F9–11–02–9D–74–E3–5B–D8–41" group on copyright issues), or Michele's specific research interests (e.g., "Machiniplex," and LOLcats). Her remaining groups are tied directly to everyday things she enjoys, like groups devoted to particular bands (e.g., Great Big Sea) or television shows (e.g., the "Addicted to Project Runway" group).

(b) Chris

Chris' formal Facebook networks containing most friends are predominantly countries and cities, including Los Angeles (where he lives), Boston, San Francisco, London and Seattle. Other formal networks include media-related or tech-related ones (Gawker Media, NTN Buzztime, and SEGA). His social networks include family and close friends, people met while traveling, friends made working within the music industry or while working as tech support in different companies and academic institutions, people met in a range of online music and discussion interest groups, and members of his current band, the ExDetectives (exdetectives.com).

At time of writing, Chris belonged to 28 groups spanning diverse interests and passions. Some are devoted to particular bands (e.g., "Numinous Eye," "Spacemen 3," "The KLF" groups), curio-cabinet type interests (e.g., "I Want to Live in the Museum of Jurassic Technology"). Others are groups devoted to musical instruments and allied gear (e.g., "VOX AC30," "Fender Guitars > all other instruments," "Pedalboard evolution"), art and moving images (e.g., "Surrealism", "Experimental Films/Avante-Garde"), politics and news critiques (e.g., "Situationist International," "Disinformation"), technology and tech news (e.g., "2600—The Hacker Quarterly"), food (e.g., "I love hickory burgers," "Chowhounds"), online affinity groups outside Facebook (e.g., "I Love Music," "Terrascope Online"), groups devoted to protesting inane copyright restrictions (e.g., "09-F9–11–02–9D-74-E3–5B-D8–41"), and groups focusing on personal habits (e.g., "People who don't sleep enough because they stay up late for no reason").

(c) Comparison

On Facebook, Michele largely defines herself in terms of her worklife. Her networks include a noticeable number of "university" friends and membership in academic groups. She uses Facebook affordances to engage in a range of social practices to do with academic work and "being an academic": for example, becoming actively involved in professional conferences, belonging to discussion and project-oriented groups in areas that match her own research interests and that extend her professional knowledge, reading others' mini feeds to keep up-to-date with work they've recently published or made available online (e.g., white papers, research reports, doctoral theses, course syllabi) that could become useful resources in her own work, and so on.

Chris' networks span diverse interests while focusing particularly on his musical interests. A guitarist in a band and with a passion for alt rock music, he engages actively in Facebook groups devoted to rock musicians (e.g., a Fender guitar group, an effects pedal group). Such participation is his way of keeping up to date with technical developments in electric guitars and establishing a presence as (among other things) a knowledgeable guitar player within the Facebook universe. His informal social networks—which span Los Angeles, New York and London—help keep him in touch with new underground bands and songs recently released outside the broadcast mainstream. In short, the networks constituting Chris's Facebook profile mediate social practices associated with being a committed aficionado and connoisseur of alt rock music styles and sounds and as a member of an alt rock band.

Michele and Chris's networks overlap around copyright issues, reflecting their respective interests in digital technologies and issues concerning overreaching and constraining copyright laws that interfere with healthy cultural development and the distribution of ideas, resources, and knowledge. In short, their respective network memberships help constitute them in socially identifiable ways within the Facebook universe.

The Textual Life of Networks

(a) Michele

Types of encodifications employed on Facebook profile pages are many and varied. Different kinds of encoded texts found on Michele's profile include images (photographs, comics, icons), video clips posted from YouTube and elsewhere, written texts in the forms of notes and written by Michele (typically including hyperlinks), status updates, wall posts made by others, and text-and-icon mini feeds. Most "textual life" within Facebook is generated through updating one's status and profile page, interacting with others directly via private messages and public wall posts, posting messages to groups, posting notes or announcements for people in one's network to read, or participating in add-on applications available to Facebook members (e.g., playing Scrabulous, taking quizzes and comparing scores with those of your friends).

We focus here on four key textual practices evident within Michele's Facebook network: the Superpoke application, wall posts, status updates, and "how we know each other" information.

Superpoke. The Superpoke application is popular among Michele's Facebook friends. This is a riff on the default "poke" application built into Facebook that lets you send a "poke" to someone with minimal loss of face should they not reply (e.g., browsing through Facebook profiles you come across a name and image that might be an old school friend; you're not sure, so you "poke" them to see if they reply with a request for Friendship). "Pokes" can also be used like a physical tap on the shoulder to remind someone that you haven't heard from them for a while. The Superpoke application is an evolved version of this. You can "throw" a range of animals and objects at others, or engage them in a range of actions (e.g., use the force on, dance with, taze). "Superpokes" appear as messages accompanied by humorous icons in your mini feed.

Sending a superpoke is easy. Michele throws a bunch of sheep at friends by clicking on the Superpoke application, selecting recipients' names from her friends list, choosing the sheep poke, and hitting send. The superpoke arrives with the option to "superpoke back," further facilitating poke exchanges by automating the process.

Wall Posts. As previously noted, walls are spaces within your profile where friends can leave short messages that can be read by everyone within that person's social network. Wall posts are an interesting textual phenomenon in that they often resemble conversations—with one half of the conversation visible on each interlocutor's page. Most Facebook users reply to messages and questions left on their wall by writing on the message sender's wall. Facebook profile readers soon learn that wall posts will often read as though they've stumbled into the middle of an ongoing conversation.

A complete wall post from one of Michele's friends, Sarah, reads: "Thanks! Yah, it's definitely a good story. :) And Reykjavik at New Years is amazing, I highly suggest it for next New Year!" The lack of contextualizing details in Sarah's post indicates that Michele earlier had posted some message on Sarah's wall. The public-ness of Sarah's response confirms that Michele's earlier message was also public, not a private message. Friends interested in following a wall conversation can click on the "wall-to-wall" option provided by Facebook, which lists the conversational exchange in chronological order. This requires users to understand the significance of the "wall-to-wall" option and take an additional step to collate a given conversation.

In Michele's social networks, wall posts within the default Facebook application typically comprise written text messages. Additional wall applications like Super Wall and Fun Wall let Facebook users post multimedia messages to friends' profile. Michele's friends post video messages filmed using webcams and embed popular YouTube video clips within a post. They also post still photos, song clips, and hand-made messages using electronic "draw" and "paint" features. Within Michele's social networks these multimedia walls typically include posted content that is humorous, related to new technology use in some way, or comprises multimedia versions of email chain letters (or spoofs of same).

For example, a recurring wallpost on Michele's profile is "Mortimer the Travelling Bear"—an image of a teddy bear accompanied by the text "Mortimer the travelling bear would like to travel all over the world. If he visits you please send him on!" Selecting the "Forward" option on this post means you can forward Mortimer to some or all of your friends in just a few clicks, with no need to add an additional message to the post. Specific functions built directly into wall-type applications are designed to facilitate social exchanges. The inclusion of simple-to-use forwarding services within wall applications makes it easy to post messages, images, videos and the like to multiple friends' walls with just a few clicks.

Wall owners can control posts to some extent by deletion. Michele's explicit orientation towards Facebook as a "professional space" means she has deleted some wall posts from friends that don't match her personal views on politics, gender, race, etc. or her projected identity as a certain kind of educator.

Status Updates. The Facebook profile menu lets you keep up to date with the current status of all friends quickly and easily in terms of what they are doing, where they are, or how they're currently feeling. Through status updates Michele has learned of friends becoming engaged to be married, deaths in friends' families, friends' upcoming television appearances or live band performances, relationship breakdowns or troubles at home, who is sick and who is recovering, who is on holiday and who is procrastinating about some task. Updates sometimes elicit messages or wall posts from others, but generally there is little expectation that friends will respond to status updates.

Michele's friends also play with or spoof the "received" purpose of status updates. When status updates by default took the form, "X is . . ." one friend insisted on using the "is" existentially, rather than in process terms, writing updates along the lines of "Tere is a stalk of grass." Typically, however, status updates lend a sense of immediacy or "now-ness" to each social network on Facebook.

"How we know each other." Facebook automates much of the "relational" information available on each user's profile, such as the descriptor categories for how friends know each other. Whenever a friend is formally added to a user's profile s/he has the option to declare how they know each other. Facebook's relationship checklist includes knowing each other through work, through previous romantic attachment, by attending school together, via taking a class or course together, via family ties, through a group or club, through travel, through a mutual friend, meeting randomly, and "I don't even know this person." Placing a checkmark beside a relationship descriptor automatically inserts it into your friend list.

These options typically include space for writing more descriptive explanations of your relationships. Within Michele's network this is often used to spoof the categories of relationship types provided by Facebook. Hence, John writes:

> You met randomly: I was trading in some of my handguns for cups of warm soup at the Bethany Lutheran Mission on 103rd Street.
>
> As I reluctantly exchanged a Walther PPK for a large bowl of cabbage and pumpkin, Michele muscled in. Speaking in German, which I immediately recognised as "enigma code," she implored me to keep the weapon. She then asked me to meet outside the (then) East German Embassy at midnight.
>
> I complied, spawning a 17 year relationship wherein we only addressed each other as "Verlaine" and "Rimbaud." Many secrets changed hands, none more important than the Stasi's list of "known harpsichordists" in Von Karajan's Philharmonic.

> The ice off the cold war has since thawed, but we still meet every winter under the
> Dutch Elm in East Berlin's Wehrmacht Park and exchange blintz recipes.
> And the rest, as they say, is history.

Facebook's option for including details about a relationship is often taken by Michele and her friends as an invitation to produce elaborate and far-fetched accounts of otherwise ordinary relationships (John and Michele met in high school and took classes in German together). Within Michele's social network such accounts tend to follow certain rules or norms, including elements of truth within each account (e.g., speaking German, attending a Lutheran high school), adopting adventure or thriller narrative styles, and grounding the account in some kind of bizarre situation (other relationship accounts include having to cut holes in ceilings to escape hordes of angry customers, competing in a Eurovision song contest, selling Mardi Gras costumes in eastern Mexico), written about in a nonchalant manner, as though such things happen to everyone. This kind of narrative work strengthens social relationships within Michele's network by establishing shared insider "jokes" and poking gentle fun at the relationship categories assumed by Facebook programmers to be significant.

(b) Chris

Chris' use of Facebook affordances differs from Michele's in significant ways, notwithstanding the fact that Facebook users share common profile architecture features. We focus here on three of Chris' characteristic practices on Facebook which, in conjunction with Michele's prevalent "ways" with Facebook will help illustrate variety in respective "realizations" of participation in social network sites as digital literacy. These practices are Chris' collocation of other online spaces within his Facebook profile, his participation in quizzes, and his membership in Facebook groups.

Collocation of Online Spaces. A self-professed "database wrangler and roving Mac Jedi," Chris has long had a dynamic online presence: personal websites in the 1990s, a highly active blog in the 2000s (http://www.quartzcity.net), and a pro Flickr account since 2002. He also uses the online bookmarking service, Del.icio.us, to record personally notable websites. Applications have been developed for Facebook users to embed feeds to blogs, Flickr accounts and Del.icio.us accounts. Chris has all three installed within his profile. These applications are completely automated; Chris simply goes about his everyday life online, posting to his blog, adding bookmarks to Del.icio.us, and uploading photos to Flickr, and each event is logged simultaneously on his Facebook profile. This makes it easy for friends to remain up to date with Chris' interests and doings, his online "finds," and his travel and food photography. Facebook's mini feed makes it even easier for friends to keep up with Chris by listing recent posts in chronological order (see Figure 11.2).

Figure 11.2: Clip of Chris' Mini Feed, Showing Blog Feeds and Del.icio. us links.

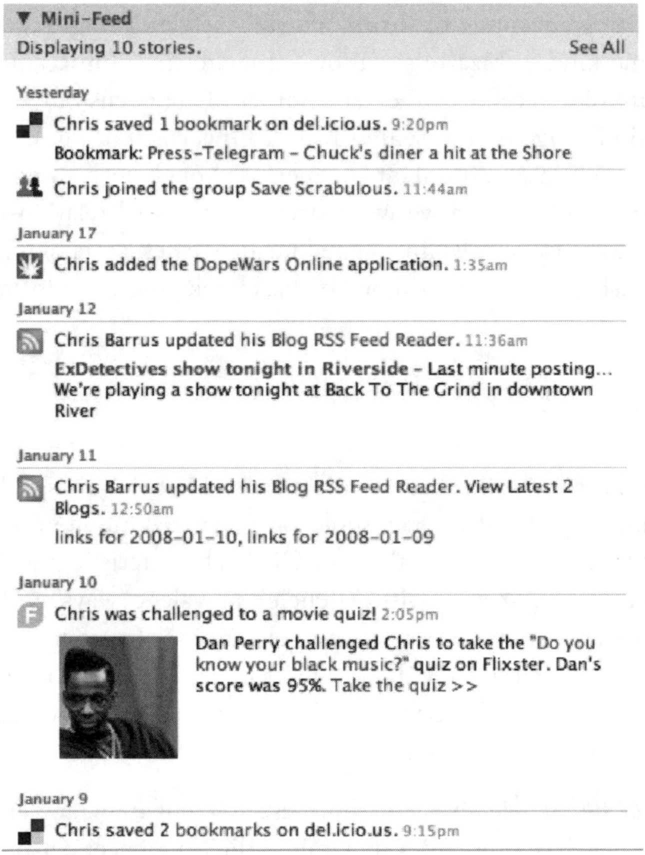

▼ Mini-Feed

Displaying 10 stories.   See All

Yesterday

■ Chris saved 1 bookmark on del.icio.us. 9:20pm
Bookmark: Press-Telegram - Chuck's diner a hit at the Shore

👥 Chris joined the group Save Scrabulous. 11:44am

January 17

Chris added the DopeWars Online application. 1:35am

January 12

📶 Chris Barrus updated his Blog RSS Feed Reader. 11:36am
ExDetectives show tonight in Riverside – Last minute posting...
We're playing a show tonight at Back To The Grind in downtown River

January 11

📶 Chris Barrus updated his Blog RSS Feed Reader. View Latest 2
Blogs. 12:50am
links for 2008-01-10, links for 2008-01-09

January 10

Chris was challenged to a movie quiz! 2:05pm
Dan Perry challenged Chris to take the "Do you know your black music?" quiz on Flixster. Dan's score was 95%. Take the quiz >>

January 9

■ Chris saved 2 bookmarks on del.icio.us. 9:15pm

(Facebook © 2008)

As Figure 11.2 shows, a typical mini feed on Chris' page includes information like the following, selected randomly over 2 days:

- "Chris saved one bookmark on del.icio.us. 9:20pm
    - Bookmark: Press—Telegram—Chuck's Diner a Hit at the

Shore [hyperlink to website]"
- "Chris Barrus updated his BlogRSS Feed Reader. 11:36am
  - ExDetectives show tonight in Riverside [hyperlink to blog post]. Last minute posting . . . We're playing a show tonight at Back to the Grind in downtown Riverside"
- "Chris Barrus updated his BlogRSS Feed Reader. View Latest 2 Blogs. 12:50am
  - links for 2008–01–10, links for 2008–01–09 [hyperlinks to blog posts]"
- "Chris saved 2 bookmarks on del.icio.us. 9:15pm"
  - Bookmarks: Anthony Bourdain | The A.V. Club, Tesla Slept Here [hyperlink to website]
  - Chris saved 1 bookmark on del.icio.us. 1:35pm
  - Bookmark: A.V. Club Taste Test Special: The Bowl at the Howling Rim of Famous-Ity [hyperlink to website]"

Such blog post and bookmark summaries inform friends widely about Chris' interests. They are notified of a live gig that the ExDetectives are playing; advised that Chris has posted two new sets of hyperlinks to his blog (for 9 and 10 January, 2008, respectively); alerted to an eating establishment in Long Beach that's worth visiting; and provided with links to an online interview with a world-famous chef, to an article about the New Yorker Hotel and one of its famous guests (Nikola Tesla) and to an hilarious review of a Kentucky Fried Chicken meal.

Facebook's architecture and add-on applications mean Chris needn't use HTML codes to insert hyperlinks within his profile. Encoding is done automatically. Clicking on entries in either the application boxes for blog feeds, Del.icio.us, or Flickr, or on links within his mini feed takes you to his blog, to sites he has bookmarked, or to his Flickr albums. Friends reading Chris' profile know the listings within the application boxes and his mini feed are only summaries of, or links to, larger texts and online spaces, and can follow up on anything catching their interest by clicking the automatically generated hyperlinks. In this way, Chris' Facebook profile acts as a portal for friends to use for following his life, rather than as a self-contained, Facebook-only social space.

Quizzes. Chris' Facebook profile indicates much about who he is in the world. One way this becomes apparent is via Facebook quizzes he chooses to take. Quiz applications he has added to his profile include movie quizzes and esoteric quizzes about non-mainstream music. He invariably scores high on the quizzes he takes, indicating a seemingly encyclopedic knowledge of popular culture and alt music scenes. His mini feed reported:

- GRUNGE ROCK B-LIST CHALLENGE!—Chris answered 10 of 10 questions correctly for a score of 100%.
- Vintage New Wave Challenge—Chris answered 14 of 14 questions correctly for a score of 100%.

To date he has never taken quizzes focusing on mainstream music. Movie quizzes he elects to complete privilege science fiction, shlock horror, music-in-films, and art film genres.

Membership of Facebook Groups.

Chris' Facebook group memberships span music interests, existential philosophies, fine arts, urban archeology and architecture, geek interests (e.g., important figures in the history of technology, programming news), and music. The group memberships appearing in his profile cohere with his blog posts, Del.icio.us bookmarks, and Flickr photos. The latter include, for example, his own photos of tunnels beneath a local university that aren't generally accessible to the public.

While posting useful links, commentaries, or clarifications via wall posts and notes contribute directly to the "life" of these groups to which he belongs, Chris' posts also convey a sense of his own expertise within, say, the rock music universe. By way of introducing himself to the Fender Guitar group, he posted the following:

Currently play:
    1965 Jazzmaster (coral)
    1975 Telecaster (blonde)
    1996 Stratocaster (black. US built standard)
    2001 Stratocaster XII (sunburst. Japanese built)

This is not a typical "Hi, my name is . . ." self-introduction to a group. Chris' list was accompanied by no commentary at all. It nonetheless conveys significant information to savvy electric guitar players. Owning four Fender guitars alone is worthy of respect among rock musicians.

Chris doesn't participate *actively* in all the groups he belongs to. Nevertheless, merely joining and listing certain groups via his Facebook profile page conveys plenty of information about his interests and habits. Membership of "Friends don't let friends use bad fonts" informs Chris' Facebook friends that he is interested in good design, good fonts, and has a definite sense of humor. Membership of the "Afflicted with effect pedal acquisition syndrome" group—mostly comprising photos of impressive collections of guitarists' effect pedals—intimates that Chris may suffer from this same "syndrome." To date he has not posted to either of these groups.

Realizations of Social Networking as Digital Literacy Practices and of Digital Literacy as Social Networking

Literacies come "whole" and any attempt to dissect them into constitutive elements runs the risk of distorting the seamlessness and intricacy of literacy practices. Nonetheless, in order to try and analyze and discuss how Michele and Chris *realize* social networking as varying enactments of a recognizable "kind" of digital literacy we will consider data from their respective Facebook profiles in terms of the analytic components of our definition of literacies, trying to keep our account as "integrated" and "whole" as possible.

Socially Recognized Ways

At a general level, signing up for and participating in a social network site is the most obvious socially recognized way of "doing" social networking as a digital literacy. As noted earlier, individuals have genuine options available here between competing services (Facebook, MySpace, Orkut, etc.), and the choices individuals make—including participating in multiple sites—reflect their beliefs about which services are best suited to supporting the kinds of networks they want to develop and participate in as "network centers" and how they want to present themselves and be perceived by others.

More specifically, however, within a particular site like Facebook participants can choose among diverse socially recognized ways afforded by the site for accomplishing self-identity presentation and interaction with friends by generating, communicating and negotiating meanings with others. Membership and active participation in groups on Facebook comprise socially recognized ways of signaling interest in, or commitment to, some particular thing. Similarly, quizzes are popular applications on Facebook serving as socially recognized ways of presenting oneself as a particular kind of person in terms of what one knows about media trivia, popular culture, general knowledge, world geography, word semantics and the like. Knowledge displays encode more than just end-point scores; they are also read as part of a person's identity since they signal personal interests and investments. Chris himself affirms that alternative, non-mainstream media productions are important elements in his life, and these interests are shared with many of his Facebook friends—as captured by their shared activity in completing esoteric alt music quizzes.

Chris and Michele both "realize" their Facebook networks by making use of default features of the Facebook profile "page" designed to support and encourage social connections. They both accept and offer friendship to people they find or who find them on Facebook. They both join Facebook groups that resonate with their respective interests. They update their status fairly regularly. Their mini feeds enable their networked friends to keep up to date with what they have been doing within the Facebook universe (and beyond). At the same time, despite the "sameness" of the look of their profile pages and their use of Facebook's default functions and services, the ways Chris and Michele use these functions vary quite markedly. In addition to using the various default functions found within the Facebook profile page, both Chris and Michele also have added a range of applications developed by third parties (e.g., games, quizzes, travel maps). These added applications serve to "customize" their profiles and provide interesting insights into how they have each chosen to represent themselves on Facebook.

Meaningful Content

One way of thinking about social networks in light of the data is in terms of "webs of insiders." A user's friends will cluster around certain interests and ways and, as in the cases of Michele and Chris themselves, there will be overlaps. The "glue" in each personal network is what is shared in common among those who "cluster" at particular points or intersections in that network. The more potential there is for commonalities to overlap—or be generated—across groups or clusters, the more potential there is for new friendships and affiliations to be made, which strengthens and develops each personal network overall. All of this, however, begins from what is shared in common between the network center (the user whose profile-driven social network it "is") and the friends who constitute formal groups or less formal clusters within this. This fact about "insiderliness" has important implications for what constitutes *meaningful content* within a network and for how meanings get made from what participants encode and how they encode it.

As noted, Facebook symbol systems affordances go far beyond conventional alphabetic text to include colored icons, photographic images, moving images, audio clips, pixel-drawn images, layout features, etc. These systems are put to diverse uses. The range of modes of expression available to users and the ease with which they can be shared across profiles mean that participants can convey multiple meanings and levels of meaning simultaneously. The exchange of "superpokes" is a good example. At one level of meaning someone who sends a superpoke is communicating "I get this and am participating because I find it fun/meaningful/whatever and I think you will too." A superpoke that throws a sheep at the recipient has a literal meaning, although it is nonsensical (as the literal meanings of many superpokes are). At another—usually the most important—level of meaning, sending the superpoke means "I'm thinking of you." Superpokes typically carry very little meaning content that is literal or has "information" value in the classic sense. Meanings made from superpoke "texts" are often almost entirely *relational*. They express solidarity, affinity, or some kind of relationship with particular other people.

The same holds for other icon-related applications (e.g., "gifts" that are small pixel-based icons sent to friends). A red stiletto shoe sent to Michele is a way of saying "I know you and this signifies something about you and I know you will relate to it. By the way, I'm thinking of you." It is not about literal giving. Of course, meaning-making that focuses on relating rather than informing or "literality" is nothing new. Nonetheless, the scale on which relational meanings are made available *and used* within social network sites like Facebook is unprecedented and has important implications for any account of digital literacy practices. Such meaning-making affordances and socially recognized ways of using them affirm Scribner and Cole's claim that literacy comprises much more than simply reading and writing and requires necessary attention to specific purposes and contexts of use (see Scribner & Cole, 1981).

With respect to encoding plain text, the example of Chris simply listing to the Fender Guitar group the guitars he plays carries much more meaning than literally meets the eye. Owning Fender models from key decades marks Chris as a connoisseur who, for example, appreciates the tonal qualities of different models. Including the year, color and country of origin of each model signals the rarity, current value (e.g., a blonde 1975 Fender Telecaster currently retails for between $5,000 and $6,000 USD), and important characteristics (e.g., the Stratocaster XII is valued for its thin neck and the complexity of sound this can produce). Listing such details also hails other guitarists who play similar guitars (e.g., Robert Smith of The Cure has played a Fender Jazzmaster in numerous live shows; Eric Clapton's guitars of choice have been Fender Telecasters, Jazzmasters, and Stratocasters). Vintage Fender guitars also carry a very high "cool quotient" for musicians. This and further meaningful content is encoded in Chris' cryptic self-introduction to the Fender guitar group. This "introduction" also suggests Chris expects that the group itself comprises savvy Fender guitar players who easily will be able to "read" the considerable information built into his list.

Much playfulness is evident in the exchanges of meaning within Chris and Michele's Facebook profiles. Despite template constraints, there is room for experimentation and humorous uses of Facebook's otherwise staid applications. This is evident in particular "spins" on the "how we met" description protocol occurring within Michele's network. Spoofing the default categories (we met at: college, high school, work, etc.) by writing short, dramatic tales of high adventure has become a shared "insider joke" within her network of friends. The "truth value" of the information contained within each relationship account is minimal and inversely proportional to the humor value her network of friends attaches to these accounts (the more an account makes people in the network laugh, the better). To be fully appreciated, however, these humorous narratives need to be read within the context of the default categories supplied by Facebook, and the purpose these "how we met" descriptions are meant to play within Facebook. In Chris' case, membership in groups like "People who don't sleep enough because they stay up late for no reason" signals a habit of his that is well known to his friends. The title and purpose of the group is amusing in itself, and its focus on a personal habit plays with traditional group charters that tend to construct groups around specific shared interests or social purposes. These kinds of playful practices emphasize the "social-ness" of Facebook and speak to the value placed on sharing jokes and spoofing certain conventions within Facebook itself.

"Getting" the different (kinds of meanings), and even attending at all to particular "texts" on someone's Facebook profile or sending certain kinds of communications to particular friends, is deeply linked to "insiderliness." Not all Chris' friends will explore the musician-oriented groups to which he belongs, just as not everyone in his network will challenge him to a movie quiz. From the standpoint of text production, Michele knows she wouldn't send a superpoke to everyone in her network because some friends would find it—or her!—"silly" or "meaningless."

Discourse Membership

The data from Chris and Michele's profiles indicate the extent to which both interact with friends from the standpoint of identifiable discursive affiliations.

This is, of course, integral to the "insiderliness" that enables meaning making of the kinds discussed briefly above. We read and write out of discursive positionings that enable meanings to be made. Different discursive positionings enable different meanings and, in some cases, may obviate making any sense of particular encodifications or may engender lack of interest in or sympathy for certain dimensions of a Facebook profile.

Michele's profile emphasizes her membership in an academic Discourse encompassing the sociocultural study of literacies and digital technologies. Her notes, group memberships and blog feed most clearly communicate this discursive affiliation. Some announcements—sent to everyone on her Facebook friends list—focus on academic writing practices. Three recent announcements included: a general call to her Facebook friends for concrete examples of teachers using remix principles and practices in their classrooms which Michele sought for a journal article she was writing; a call for book reviewers in her role as Book Reviews Editor for an academic journal; and a conference announcement by the Canadian Games Studies Association, originally sent to her via email. These examples suggest Michele is actively shaping and being shaped by an academic Discourse that requires members to publish their work (and which requires concrete examples of evidence from classrooms to support claims rather than, say, arguing in more *a priori* ways as one might in philosophy). This same Discourse means that when friends read her call for book reviewers, everyone appeared to realize that the books being offered for review were academic texts, and not, say, novels (not one of her friends working outside universities expressed interest in any of the books on offer). The announcement about the video games conference was sent directly to Michele from a member of the Canadian Games Association, suggesting that Michele is considered by the sender to have networks of distribution comprising people interested in the conference announcement.

Michele's use of notes to extend her academic Discourse practices into her Facebook networks likewise appears to be understood by her friends. Hence, even though many of her friends are active video games researchers, none replied to her announcement about the games conference, seemingly recognizing that the purpose of the message was simply to inform others about the conference. Anyone interested in the conference knew to click on the hyperlinks within her announcement and follow up on conference details of their own accord. This contrasted markedly with her call for book reviewers; not only did friends reply immediately via the private messaging function on Facebook, nominating which books they were interested in reviewing, but friends also passed this note along to others (e.g., Facebook friends' doctoral students got in touch about reviewing, using conventional email). This comparison of responses underscores how insiders to a Discourse seem to know "automatically" and collectively how to make sense of different texts. The Facebook groups Michele has joined mainly focus on digital technologies, literacy and education. Her membership in them enables her Facebook friends—even those who are not themselves academics or educators—to "recognize" her as having a particular kind of identity as an academic working within the field of literacy and digital technologies.

The most visible discourse co-ordinations (Gee, 1997) on Chris' Facebook profile belong to an alternative or "alt rock" Discourse. As Chris's textual work and social practices indicate, full membership in this Discourse entails much more than merely listening to favorite music. It also involves Chris in reading about bands and the history of certain musicians, music periods, or genres; participating in bands himself as a musician; having an online presence within a range of alt music discussion boards; and knowing a good deal about rock-related musical instruments and gear. For example, Chris' Virtual Bookshelf includes titles like *Visual Music: Synaesthesia in Art and Music Since 1900* by Ari Wiseman, *Lollipop Lounge: Memoirs of a Rock and Roll Refugee* by Genya Ravan, and *Rip It Up and Start Again: Postpunk 1978–1984* by Simon Reynolds, to name a few. He has installed the "What I'm Listening To" application within his profile, which synchs with his computer-based music player. Whatever he's listening to while working at his computer appears automatically on his Facebook profile. A recent listing of songs captures the eclectic range of music in which he's interested: "Stone Lost Child" by Lee Hazlewood, "Love Is Not Real" by The Negatives, "Glory Bee" by Lightnin' Hopkins, "Transition Man" by The Launderettes, and "Whiskey Rebellion" by Econoghost. These span blues (which continues to have a deep and pervasive influence on alt rock), punk rock, garage punk, country rhythm and blues, and their origins range across southern U.S., Norway, England, and Germany.

Chris' group memberships also speak to his "insider" status as an alt rock musician and aficionado. His Flickr feed within his profile further supports this, with his most recent photos showing the ExDetectives playing at Hollywood's Knitting Factory. The quizzes he completes often focus on music (e.g., the Vintage New Wave challenge, the Grunge Rock B music challenge), and he rarely scores below 100 percent accuracy. Further evidence of his "insider" status within the alt rock Discourse is evident in his list of Facebook friends, many of whom, he explains, he met through online discussion forums devoted to non-mainstream music. While Chris' Facebook profile is interesting in its own right it is especially so to people who share his participation in the alt rock Discourse.

Encodifications

Some of the most interesting things about the kinds of texts encoded in social networking sites and the means available for coding them concern the ease of encoding and the relative absence of extended codifications of alphabetic text. Social networking "work" gets done by means of encodification that is significantly different from more familiar literacy practices in physical-print space (e.g., letter writing) as well as in digital media spaces like weblogs, email clients, conventional websites, and so on.

It is easy to participate within Facebook's social networks without knowing a great deal about hypertext mark-up language or other programming scripts. Moreover, because so much of the encoding needed to display texts online is automated within Facebook, it is possible for users to send each other complex texts like video clips, sound files and images without having to hand-code the interface display, worry about internet protocols for storing and sending bits and bytes, and so on. This ease of use may well explain why many of the texts exchanged within Chris' and Michele's networks of friends tend to be visual, rather than alphabetic, in nature. This practice is very much in keeping with a range of sociocultural commentaries on digital literacies that describe the rising dominance of visual modes of meaning-making over written language (Cope et al., 2005; Lessig, 2005; Perkel, in press). The emphasis on "ease of use" within Facebook is exemplified especially well in applications that can be installed in your profile and that enable you to display updates and summaries of what you are doing elsewhere on the internet. Chris' mini feed is a typical case: with just a few mouse clicks he can display on his profile page summaries of recent blog posts, Flickr posts, and his online bookmarks, along with samples of the music to which he is currently listening.

Conclusion

Construed as digital literacy *practice*, social networking pursues "socially recognized goals" by means of using "a shared technology and knowledge system" (Scribner & Cole, 1981, p. 236). Goals include presenting and constituting oneself as a particular kind of person; performing an identity that is partly defined by the friendship network that develops around each user at the center of his or her network. This presentation and constitution of self is largely enacted in the choices (typically without conscious attention to identity "presentation") each person makes about what is displayed on their profile page, what they post on or send to or engage in via other people's profile pages (e.g., sending superpokes, challenging a friend to a quiz), what gets posted to their own page, and the interaction that takes place with others in their network via a range of Facebook applications.

In the case of Facebook, the network technology itself shapes the diverse affordances of the profile architecture, including the mini feed, wall posts, the range of opportunities to compare oneself with others through quiz results, game scores, and the like. The knowledge system involved in participating within Facebook is complex. Part of it involves "co-ordinated sets of actions" or "skills" (Scribner & Cole, 1981, p. 236) concerned with performing the technology—such as knowing how to update your status by clicking on the text to activate the dialogue box; knowing how to generate "how we met" statements beyond the default categories; knowing how to add (and delete) different profile applications, and so on. Beyond this, however, the knowledge systems that users bring to their social networking involve discursive knowledge and how to render this effectively by means of various modes and text types such that strong and expansive connections are made with like minds and kindred spirits by sharing meanings that mutually enrich, sustain, and expand participants' everyday lives. This may involve awareness and enactment of more or less distinctive personal styles (e.g., Chris' way of presenting himself to the Fender Guitar group, or by posting photos of underground tunnels that mark him as someone interested in urban exploration and other quirky things; or Michele's use of the notes application to post news about academic events, or the ways in which superpokes are used within her particular network).

As digital *literacy* practice, social networking involves recognized ways of using the encoding affordances of services like Facebook to generate, communicate and negotiate personally significant meanings from the standpoint of participants who come to Facebook as members of varying Discourses whose integral ways of doing, thinking, valuing, acting, believing, speaking, gesturing, appreciating, etc., constitute participants respectively as particular kinds of persons and situated selves (Gee, 2004). Thus, Chris' "text bites" convey a sense of eclectic interests in food, urban architecture, popular culture, obscure or forgotten accounts of historical events and people and, of course, music. Similarly,

Michele's text bites convey her interests in literacy and digital technology studies, in spoofing default categories made available within the profile template, and in maintaining relationships via the medium of Facebook. To this extent and in these ways we can understand participation in social networking sites from a sociocultural perspective as digital literacy practice performed in diverse ways by ordinary people who live their everyday lives to a greater or lesser extent online.

References

Baym, N. (2007). The new shape of online community: The example of Swedish independent music fandom. *First Monday*, 12 (8). Available at: http://firstmonday.org/issues/issue12_8/baym/index.html

Boyd, D. (2008). Why youth (heart) social network sites: The role of networked publics in teenage social life. In D. Buckingham (Ed.), *Youth, identity, and digital media* (pp. 119–142). Cambridge, MA: MIT Press.

Boyd, D. & Ellison, N. (2007). Social network sites: Definition, history, and scholarship. *Journal of Computer-Mediated Communication*. 13(1). Available at: http://jcmc.indiana.edu/vol13/issue1/boyd.ellison.html

Cassidy, J. (2006, May 15). Me media. *The New Yorker* (pp. 50–59).

Cope, B., Kalantzis, M., & Lankshear, C. (2005). A contemporary project: An interview. *E-Learning 2*, 2: 192–207.

diSessa, A. (2001). *Changing minds: Computers, learning, and literacy*. Cambridge, MA: MIT Press.

Ellison, N. B., Steinfield, C., & Lampe, C. (2007). The benefits of Facebook "friends": Social capital and college students' use of online social network sites. *Journal of Computer-Mediated Communication*. 12(4), article 1. http://jcmc.indiana.edu/vol12/issue4/ellison.html

Gee, J. (1997). Foreword: A discourse approach to language and literacy. In C. Lankshear, *Changing Literacies*. Buckingham: Open University Press.

———. (2004). *Situated language and learning: A critique of traditional schooling*. London: Routledge.

———. (2007). *Good video games + good learning: Collected essays on video games, learning and literacy*. New York: Peter Lang.

Humphreys, L. (2007). Mobile social networks and social practice: A case study of dodgeball. *Journal of Computer-Mediated Communication*. 13(1). Available at: http://jcmc.indiana.edu/vol13/issue1/humphreys.html

Kumar, R., Novak, J., & Tomkins, A. (2006) Structure and evolution of online social networks. *Proceedings of 12th International Conference on Knowledge Discovery in Data Mining* (KDD-2006) (pp. 611–617). New York: ACM Press. Philadelphia, Pennsylvania, August 20–23, 2006.

Lampe, C., Ellison, N., & Steinfeld, C. (2007). A familiar Face(book): Profile elements as signals in an online social network. *Proceedings of Conference on Human Factors in Computing Sys-*

tems (CHI 2007) (pp. 435–444). New York: ACM Press. San Jose, CA.

Lessig, L. (2005). Keynote address presented at the ITU "Creative Dialogues" Conference, University of Oslo, 21 October 2005.

May, M., & Kwong, K. H. (2007). *YHOO: Yahoo! may regret not paying up for Facebook.* Retrieved May 10, 2007. Available from: http://www.needhamco.com/Research/Documents/CPY25924.pdf

Paolillo, J. & Wright, E. (2005). Social network analysis on the semantic web: Techniques and challenges for visualizing FOAF. In V. Geroimenko & C. Chen (Eds.), *Visualizing the semantic web* (pp. 229–242). Berlin: Springer.

Perkel, D. (in press). Copy and paste literacy? Literacy practices in the production of a MySpace profile. In K. Drotner, H. Jensen, & K. Schroeder (Eds.), *Informal learning and digital media: Constructions, contexts, consequences.* Newcastle, UK: Cambridge Scholars Press.

Scribner, S. & Cole, M. (1981). *The psychology of literacy.* Cambridge, MA: Harvard University Press.

Snyder, J., Carpenter, D., & Slauson, G. (2006). *Myspace.com: a social networking site and social contract theory.* Dallas, TX: ISECON 23. Retrieved 19 November, 2007. Available from: http://isedj.org/isecon/2006/3333/ISECON.2006.Snyder.pdf

CHAPTER TWELVE

Digital Literacy and the Law

Remixing Elements of Lawrence Lessig's Ideal of "Free Culture"

ASSEMBLED AND REMIXED BY

COLIN LANKSHEAR AND MICHELE KNOBEL

Introductory Note

In this chapter we have assembled and lightly remixed some key ideas from three texts by Lawrence Lessig to present a perspective on how contemporary concepts, policies, and laws of copyright and intellectual property impinge on practices of digital literacy. The source texts for this particular assemblage and remix of ideas are Lessig's 2004 book *Free Culture: How Big Media Uses Technology and the Law to Lock Down Culture and Control Creativity* and two variants of his renowned illustrated oral address on the themes of copyright, free culture and Creative Commons that has been presented many times and in many places around the world.

On January 2008 Lessig announced on his blog (http://www.lessig.org/blog/) that he was giving his final public address on "Free Culture" and would henceforth be devoting his energies to addressing corruption in Washington, DC. This, then, is an opportune moment to re-present what we see as some of his key ideas and to bring them into conversation with our own particular interests in education and digital literacies.

This chapter is presented as an "assemblage" under *our* names rather than Lessig's at his request and in accordance with permission to remix selections from his work. There is no implication whatsoever that he personally endorses this chapter. We greatly appreciate his permission to publish this remix and esteem his landmark work on behalf of the ideal of free culture—in the sense of "free" advocated by Lessig and initially articulated by Richard Stallman (2002)—and the right to fair use in the interests of creative expression and cultural production (see other remixes of Lessig's words at: http://www.free-culture.cc/remixes).

What follows is *not* "our own work" in any significant sense. It represents our particular selections, editing and re-voicing of excerpts from an extant corpus of written and oral texts authored by Lessig in accordance with our choice of argument structure. To produce this chapter we have done the following things.

First, we generated verbatim transcriptions of recordings of the two talks, so far as it was possible to "catch every word." We spent a lot of time reading these transcripts, locating online as many as possible of the artifacts referred to in the talks, and consulting these artifacts.

Second, we then excerpted and arranged stretches of the transcriptions in accordance with a scope and sequence of argument we thought would meet our purposes for this book and, especially, that would provide a useful introduction to readers who may not previously have considered the kinds of issues Lessig addresses academically, professionally and politically. We have stayed as close as possible to the wordings in the transcriptions (from Lessig 2005a, 2005b) for each passage selected. Where appropriate, we have revoiced the text or reported content in the transcripts by paraphrasing. We have moved between the talks, using one talk to render a particular idea and the other talk to render another, according to our personal preferences. To a very large extent the words presented here from the talks correspond directly to passages from our verbatim transcriptions.

Third, we consulted *Free Culture: How Big Media Uses Technology and the Law to Lock Down Culture and Control Creativity* (available at http://www.free-culture.cc/freeculture.pdf) and identified passages we thought would best augment content in the talks for our purposes here.

Fourth, we then excerpted (with full citation) and paraphrased passages from pages 7–8, 36–38, and 140–144 of *Free Culture* and integrated this content into our argument structure.

Fifth, we have inserted ourselves more directly into the text in the Conclusion, where we draw briefly on some current themes in educational theory and research to distinguish a "content transmission" view of education from an ideal of "expert performance" in a range of social roles and identities. The cultural transmission model raises the stakes for "ownership of intellectual property" and, to that extent, goes hand in glove with intensified copyright legislation and use of "digital rights management" technologies—to what Lessig (2005b) calls a permission culture. The expert performance model presupposes keeping culture as free as possible, so that learners have maximum scope to become experts through acts of tinkering and remixing.

Finally, we have archived the audio recordings of Lessig's talks online, at http://www.coatepec.net/lessig2005a.mp3 (i.e., Lessig, 2005a) and http://www.coatepec.net/lessig.2005b.mp3 (i.e., Lessig, 2005b) and our verbatim transcriptions of these talks at http://www.coatepec.net/lessig2005a_transcript.pdf and http://www.coatepec.net/lessig2005b_transcript.pdf (all used with permission). This will enable anyone who wants to do so to check the veracity of the following text as well as to gauge proper attribution. To anyone who may wish to quote directly from this chapter we ask them to compare the following text with the transcriptions and so far as they deem appropriate to use a form like "Lawrence Lessing as cited/presented/remixed in Lankshear and Knobel, 2008."

Writing and Remix

We can begin by thinking about what is meant by "writing." What do we do, or imagine that we do, when we ask people—in particular, children and adolescents—to engage in writing? In the United States there has long been a practice in schools of engaging in creative writing. For example, students read a book by Hemingway, *For Whom the Bell Tolls*, and a book by F. Scott Fitzgerald, *Tender Is the Night*. Then they take bits from each of these books and put them together in an essay. We take and combine, and in such classes that's the writing—the creative writing—that constitutes education about writing (Lessig, 2005a). This is a practice of taking and remixing as a way of creating something new through writing.

From this particular standpoint literacy can be seen as the practice of teaching and learning how to remix other people's texts. Indeed, this idea of writing has often been at the core of education in our tradition. If we reflect upon its tools we recognize that some of its historically familiar tools include artifacts like pencils and typewriters. These are the physical things we use to engage in writing. The resources we use are, of course, words drawn from text. We use these words or text and these tools to make new text: that is, we *remix* text.

To introduce a point that we will come back to later, we could add that this practice of writing involves a particular way of thinking about how we learn to write. We learn to write in one simple way: by doing it. We have a literacy that comes through the practice of writing: "writing" meaning "taking these different objects and constructing or creating with them." It is through our fingers on the pencil or typewriter keys that this knowledge comes. For the purposes of this chapter, that is what writing means and how we learn to write.

Remix and Culture

As such, writing in this sense is just one part of a much more general practice of cultural engagement that, following Lessig (2005a, 2005b) we may call *remix* (see also Lankshear & Knobel, 2006, Ch. 4; Knobel & Lankshear, 2008). Remix is the idea of someone mixing cultural resources together, and then someone else coming along and remixing the thing the previous person had created, by selecting from it and adding new cultural resources to it, and inserting their own purposes and inflections into it. In this sense, we could say—broadly speaking—that culture is remix. We could say that knowledge is remix, that politics is remix, and so on. Always and everywhere this is how cultures have been made—by remixing; taking what others have created, remixing it, and sharing with other people again. This is what cultures are.

Remix is everywhere engaged in by everyone. It is something we all do and it is how we live. We watch a movie by Michael Moore and then tell our friends about how it is the best movie we have ever seen or the worst movie ever made. When we do this we are, in effect, taking Michael Moore's creativity, remixing it, and thereby involving it in our own lives. We are using it to extend our own views, or to criticize his views, or to bring his views into conversation with our own or those of other people. In doing so we are taking elements of culture and engaging in acts of remixing. In this sense every single act of reading and choosing and criticizing and praising culture is an act of remix. Moreover, it is through this general level of remixing practices that cultures get made. (Lessig 2005b)

At a different level corporations engage in remix. Apple, for example, takes the iPod and remixes it quite regularly. Vehicle-producing corporations are renowned remixers as are manufacturers of diverse kinds of new technologies.

Similarly, politicians remix. Some would say (and, indeed, many *have* said) that in 1992 Bill Clinton took the Republican Party platform, remixed it a little and, with it, became a Democrat president. Of course, remix goes across the political spectrum. In the U.S., conservatives did it during the 2004 presidential election:

> Music plays.
> Massachusetts senator John Kerry. Hair style by Christofs, $75. Designer shirt, $250. 42 foot luxury yacht, one million dollars. Four lavish mansions and beachfront estate, over thirty million dollars. Another rich liberal elitist from Massachusetts who claims he is a man of the people. Mmmmmm. Priceless. (excerpted from a visual presentation of the advertisement spoofing a Mastercard commercial in Lessig, 2005a)

Liberals do it too. At a popular website of antiwar posters (www.Anti-WarPosters.com) we find [on 20 November 2007 it was the 54th poster in the slideshow] a remixed vintage war poster depicting a man in a classic brown suit and waistcoat, sitting in an expensive-looking armchair, in a pensive pose. His young daughter is sitting on his knee, with her finger on the page of an open book, while his young son is playing soldiers on the carpet at his feet. Beneath the image the caption reads: "Daddy, why don't YOU or any of your friends from ENRON have to go to war?" (cited in Lessig, 2005a. For similar examples, see Lankshear and Knobel, 2003, p. 42, and Lankshear and Knobel, 2006, p. 134).

Digital Remix

We have earlier mentioned some familiar tools of writing—the pencil and the typewriter. These tools have now changed. We have also presented Lessig's general description of the practice of remixing. Remixing practices have changed as well. Right now the key tools of remix and of writing as remix are *digital*. The internet, particularly, is awash with examples of ways in which these digital tools get used in the practice of creativity.

One of the best-known everyday examples of digital remix is found in the burgeoning popular cultural practice of AMV (anime music video) production. AMV culture is a mix of anime, music and video culture. Anime are the various Japanese cartoon series on television. Aficionados of AMV culture record a full series of these anime cartoons and then re-edit them and synchronize them to music (Lessig, 2005a, 2005b; see also Lankshear & Knobel, 2006, Ch. 4). So we now have a kind of music video but using anime. Many of these cultural producers are kids sitting in their bedrooms with computers taking images that they capture and re-expressing them by synching them to music or to sound tracks of film trailers. Lest we think that this is the extraordinary creativity of Japanese youth it is important to recognize that these AMV creators are *not* necessarily Japanese kids. They're a community made up largely, if not still primarily, of American kids—collectively from a multiplicity of ethnic backgrounds—numbering in the hundreds of thousands, connected through the internet and engaging in this particular practice of remix. In their productions, the form of their music videos is constrained by the particular anime series or soundtrack they are using to create their images or trailer with, and the challenge is to achieve originality within the parameters of synchronizing music and animation (Lessig, 2005b).

In 2004, says Lessig, a film called *Tarnation* captured the imagination of attendees and judges at the Cannes film festival; a BBC article reported that it "wowed Cannes." This was a 2003 documentary produced by Jonathan Caouette for $218.32, using free iMovie software to remix various kinds of digitally-scanned analogue material (like photographs, home movie and VHS footage) on a Mac computer. Caouette took video from his life and put it together with a musical score he compiled from existing work by recording artists using digital technologies to produce a film that would inspire Cannes, as well as win awards for best documentary film at the Los Angeles and London international film festivals, along with several other awards.

We see it also in the context of music. Everybody knows the album by the Beatles called *The White Album*. This inspired an album by rapper Jay-Z called *The Black Album*, which in turn inspired an album by DJ Danger Mouse called *The Grey Album*. Danger Mouse mixed samples of instrumental tracks from the Beatle's *White Album* with rapper Jay-Z's acappella *Black Album* to create 12 distinctively different songs (see http://www.illegal-art.org/audio/grey.html).

As Lessig explains, the kinds of creations represented by *Tarnation* and Danger Mouse were possible in 1970, but only if you were a television station or a film or music studio producer. But now anyone with a $1500 computer can capture images and sounds from the culture around them and make sophisticated statements using these resources. It is a kind of writing, and one that is changing much in our world. It is changing the feel of society; changing the freedom of speech by changing the powers of speech. It is not merely re-creating a kind of broadcast democracy, but, rather, is increasingly building a bottom-up democracy. As Lessig (2005b) puts it, these new initiatives are not reproducing a *New York Times* kind of democracy but are engendering a network(ed) democracy. This is not a matter of the few speaking to the many, but of people increasingly speaking peer to peer. This is also changing the way we speak about writing. This is what *constitutes* writing in the early 21st century. It is writing with a different set of tools. It is the same activity but with different "words." But it is now not just words. Rather, it is words with sounds and images and video—with everything our culture consumes. All this material, digitized, is the source for this writing.

Writing in the Early 21st Century

As we increasingly understand how youth use these technologies we recognize that it is through this freedom of use that they come to know the world. They know it through their capacity to tinker with the expressions that the world gives them, in just the same way we came to know the world when we tinkered with its words. But now the world is different. Today's young people engage in active creative remixing of the culture that is around them. For them, writing is exactly what they do every time they remix the kinds of resources that they (and we) consume.

There is a very important educational dimension to this. Just as there is a grammar for the written word, so, too, is there a grammar for media. Just as young people learn how to write by writing lots of what is often at first terrible prose, so they (and we) learn how to write media by constructing lots of (what, at least, at first may be) terrible media.

A growing field of academics and activists sees this form of literacy as crucial to the next generation of culture. For although anyone who has written conventional text understands how difficult writing is—how difficult it is to sequence the story, to keep a reader's attention, to craft language to be understandable—few of us have any real sense of how difficult digital media production is. More fundamentally, perhaps, few of us may have a clear sense of how media works; of how it holds an audience or leads its audience through a story, or of how it triggers emotion or builds suspense (Lessig, 2005a).

As Lessig (2004) states, we learn to write by writing and then reflecting upon what we have written. In parallel ways, "one learns to write with images by making them and then reflecting upon what one has created" (p. 36). This grammar has changed in tandem with changes in media. When visual media were mainly just film, as Elizabeth Daley, who was executive director of the University of Southern California's Annenberg Center for Communication and Dean of the USC School of Cinema-Television at the time, explained to

Lessig, the grammar was about "the placement of objects, color, . . . rhythm, pacing, and texture" (interview with Elizabeth Daley & Stephanie Barish, 13 December, 2002, in Lessig, 2004, p. 37). That grammar changes, however, when computers open up an interactive space where a story is "played" as well as experienced. "Simple control of narrative is lost, and so other techniques are necessary" (ibid.). Lessig's example of science fiction author Michael Crichton is germane here. Crichton had mastered the narrative of science fiction, but when he turned his hand to designing a computer game based on one of his works it was a different matter. This was "a new craft he had to learn" (ibid.). Even to "a wildly successful author" it was not obvious how "to lead people through a game without them feeling they have been led" (ibid.).

The push for an expanded literacy is not, however, about making better film directors. Instead, as Daley explained (in Lessig, 2004, p. 37),

> probably the most important digital divide is not access to a box. It's the ability to be empowered with the language that that box works in. Otherwise only a very few people can write with this language, and all the rest of us are reduced to being read-only.

Lessig identifies "read-only" as the twentieth century world of media and welcomes the way in which the twenty-first century could be different, by being both reading *and* writing. At the very least it might be "reading and better understanding the craft of writing" (Lessig, 2004, p. 37). But at best it would be "reading and understanding the tools that enable the writing to lead or mislead." From this perspective, the aim of any literacy, and the aim of the kind of media literacy that Daley, Barish and Lessig—among legions of others—are concerned about in particular, is to "empower people to choose the appropriate language for what they need to create or express" (Lessig, in interview with Daley & Barish, 2004, p. 37). It is to enable students "to communicate in the language of the twenty-first century" (ibid.).

In an interesting paradox, educators often find that extending to young people opportunities to learn to write with the new tools of writing can become "a powerful and productive conduit for learning to write (or write better) with 'the old words'" (Lessig, 2004, p. 37). Interestingly, the language of digital media comes more easily to some young people than it does to others—just as is the case with any kind of language. And, not surprisingly, perhaps, it does not necessarily come more easily to those people who excel in formal written language. From this standpoint it is worth noting the potential that providing formal learning opportunities in writing as remix might have for redressing traditional inequities within formal education. Leaving such considerations aside for the present, however, Daley and Barish reported to Lessig an especially poignant case of a project they ran in a low-income area, inner-city Los Angeles school. By all conventional measures of success, this school was a failure. Daley and Barish's program gave learners an opportunity to use film to express meaning about something they knew about: namely, gun violence.

The account of the class relayed by Barish affirms an increasingly familiar experience of educators who address challenges of literacy education at least in part by shifting activity onto ground that includes some of the new tools and practices of writing: notably, digital remix. Familiar problems were turned on their heads. The Friday afternoon class presented the school with a new kind of challenge. In place of the usual challenge—to get the kids to come to class—the new class presented the challenge of keeping them *out* of class. Barish described learners "showing up at 6 A.M. and leaving at 5 at night" (Lessig, in interview with Daley & Barish, 2004, p. 38). The students were working harder than in any other class to do what education should be about: namely, promoting learning on the part of young people about how to express themselves.

Barish described the class using whatever "free web stuff they could find" in conjunction with relatively simple tools that enabled the students to mix "image, sound, and text." Participants produced a series of multimedia projects that communicated information and meanings about gun violence that few people who live outside such experiences on a regular basis would otherwise understand. Gun violence was an issue that was close to the students' lives and, as Barish explained, the project "gave them a tool and empowered them to be able to both understand it and talk about it" (ibid.). The tool made available in the project created a medium for the students to engage in self-expression. And it did this much more successfully and powerfully than could possibly have occurred if alphabetic text had been the only resource available. Barish claimed that "[i]f you had said to these students, 'you have to do it in text,' they would've just thrown their hands up and gone and done something else" (cited in Lessig, 2004, p. 38), partly "because expressing themselves in text is not something these students can do well" (ibid.). At the same time, of course, text is not a medium in which the kinds of ideas in question can be expressed particularly well. Much of the power of the messages the students in this class were successful in conveying about gun violence resulted from being able to connect the understandings and experiences to the form of expression made available through the technologies of digital remix.

Writing and Freedom

For young people writing is exactly what they do every time they remix the digital resources of everyday consumption. This raises the question: Is this writing allowed?

Of course writing is allowed in the old sense of writing; that is, with words. We have long been free to engage in the form of writing that remixes short passages of Ernest Hemingway with short passages of F. Scott Fitzgerald. No one would ever question or doubt the freedom built into that creative critical activity. But what about this new form of writing; using digital technology and images from the culture around it? Is this form of writing allowed? We may ask, more broadly, when the tools of writing change, do freedoms change as well? And is remix as free for today's children—using the tools that they have to remix cultural resources with—as it was for us of previous generations? Or to put it another way again, if "writing" is allowed for those of us over 40, in the sense of writing that we grew up with, will writing be allowed for them, in the sense that *they* are growing up with?

Remixing culture teaches as well as creates. It develops talents differently, and it builds a different kind of recognition. Yet the freedom to remix these objects—including plain text—is *not* guaranteed. On the contrary, that freedom is increasingly highly contested. The law and, more and more, technology interfere with a freedom that digital technology, and curiosity, would otherwise ensure.

Yet, as John Seely Brown, a foremost learning scientist with a deep interest in creating knowledge ecologies to foster innovation, observes, "This is where education in the twenty-first century is going" (cited in Lessig, 2004, p. 47). We need to "understand how kids who grow up digital think and want to learn." But in direct opposition to this, as Brown observes, "we are building a legal system that completely suppresses the natural tendencies of today's digital kids. . . . We're building an architecture that unleashes 60 percent of the brain [and] a legal system that closes down that part of the brain" (cited ibid., p. 47).

This new kind of writing and creating needs the same freedom that the writing of the 20th and 19th and 18th centuries needed. To do it well, to understand how it works, to practice it, to teach it and to develop it—to develop the literacy that is embedded in this writing—requires the same kind of freedom the earlier writing enjoyed.

Throughout the period of modern education in societies like our own, we have needed no special permission to engage in the "ordinary ways" that everyday citizens have for engaging in remix, including the particular practice of remix that we call writing.

Hitherto, this practice has been free in our cultural tradition. The presupposition has been that this act of taking and re-including within the process of creating is free. We need no permission to engage in it. It is free to quote or to cite. It is something that legally we should be able to do freely, without needing permission first. We take it for granted that this is the permission granted to us as citizens and as authors. Historically, however, this freedom was not automatically given. Rather, it was won. It took many generations and traditions to establish, at least in the world of text, that the right to take and remix without permission was given; it was a practice that was free (Lessig, 2005b).

In Anglo-American culture, everyday "non-commercial" practices of remix have been free in the sense of being "unregulated." By "commercial culture" Lessig (2004, p. 7) means "that part of our culture that is produced and sold or produced to be sold." "Non-commercial culture" refers to everything else. Hence, "when old men sat around parks or on street corners telling stories that kids and others consumed, that was noncommercial culture. When Noah Webster published his 'Reader,' or Joel Barlow his poetry, that was commercial culture" (ibid., pp. 7–8).

Throughout the greater part of the legal tradition of the United States, non-commercial culture was essentially unregulated.

> Of course, if your stories were lewd, or if your song disturbed the peace, then the law might intervene. But the law was never directly concerned with the creation or spread of this form of culture, and it left this culture "free." The ordinary ways in which ordinary individuals shared and transformed their culture—telling stories, re-enacting scenes from plays or TV, participating in fan clubs, sharing music, making tapes—were left alone by the law. (Lessig, 2004, p. 8)

In the United States the focus of the law was on *commercial* creativity. Lessig describes how "at first slightly, then quite extensively, the law protected the incentives of creators by granting them exclusive rights to their creative work, so that they could sell those exclusive rights in a commercial marketplace." (2004, p. 8) Although, as Lessig notes, this is not the sole purpose of copyright it is, however, overwhelmingly the central purpose of the copyright that was established in the federal constitution. Besides protecting the commercial interest in publication, state copyright law also historically protected a privacy interest (ibid., p. 308, note 8); "By granting authors the exclusive right to first publication, state copyright law gave authors the power to control the spread of facts about them" (ibid.). Safeguarding creator incentives is an important part of creativity and culture, and one which has become an increasingly important part of culture in North America. But Lessig notes that in no sense did this dominate the U.S. tradition historically. It was simply just one part—a controlled part—that was balanced with the free part of culture.

But this historical picture is changing rapidly and profoundly as analogue media are increasingly augmented with digital media.

The New Digital Default

Returning to the question about whether the new kind of writing is free we find that the simple answer, based on the perspective of American and European Union conceptions of the rules of copyright, is "No."

Lessig explains this is because against the background of copyright law the new uses of culture in the kinds of remix practices described above are technically illegal. The reason for this illegality stems from the fundamental digital inversion of copyright that has been produced by the digital network. The default position with respect to freedom to use and share culture has been flipped—inverted—by the architecture of digital technology (Lessig, 2005a). The default in the analogue world was that our freedom to act on and to use culture was free, in the sense of "unregulated." But the default in the digital world is *regulated culture*.

In *Free Culture: How Big Media Uses Technology and the Law to Lock Down Culture and Control Creativity* (2004, Ch. 10, pp. 140–145), Lessig uses a series of diagrams based on the case of the book—a physical object from the analogue world—to show how control over writing and creativity as remix with respect to freedom to use and share culture gets flipped as a consequence of the architecture of digital technology.

Lessig uses a circle to represent all of the potential *uses* of a book.

Figure 12.1. Potential Uses of a Book. Adapted from Lessig, 2004, p. 140.

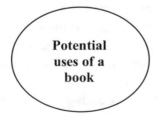

Potential
uses of a
book

As it happens, most of the potential uses of books are not regulated by copyright law because these uses do not involve making a copy. Using likely and less likely examples of uses of a book, Lessig observes that

> [i]f you read a book, that act is not regulated by copyright law. If you give someone the book, that act is not regulated by copyright law. If you resell a book, that act is not regulated (copyright law expressly states that after the first sale of a book, the copyright owner can impose no further conditions on the disposition of the book). If you sleep on the book or use it to hold up a lamp or let your puppy chew it up, those acts are not regulated by copyright law, because these acts do not make a copy. (2004, p. 141)

The following diagram summarizes some unregulated uses of a book.

Figure 12.2. Unregulated Uses of a Book. Adapted from Lessig, 2004, p. 141.

Of course there *are* uses of a copyrighted book that are regulated by copyright law. These include republishing the book, because this makes a copy. Accordingly, republishing a book is regulated by copyright law. Lessig identifies this particular use as standing at the core of a circle of possible uses of a copyrighted work. Republishing "is the paradigmatic use properly regulated by copyright regulation" (2004, p. 141)—as represented in the following diagram.

Finally, in addition to unregulated and copyright regulated uses of a book "there is a tiny sliver of otherwise regulated copying uses that remain unregulated because the law considers these 'fair uses'" (Lessig, 2004, p. 142).

Figure 12.3. Unregulated Uses Compared with Regulated Uses of a Book. Adapted from Lessig, 2004, p. 142.

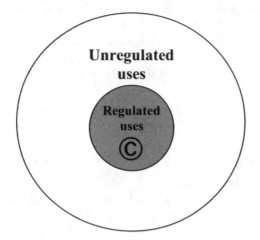

Figure 12.4. Fair Use Compared with Regulated and Unregulated Uses of a Book. Adapted from Lessig, 2004, p. 142.

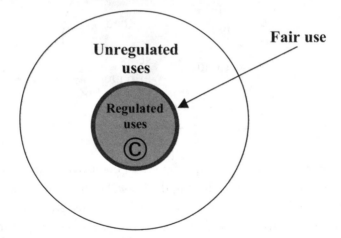

These "fair uses" are uses that involve making a copy, but the law treats them in the manner of unregulated uses because public policy requires them to remain unregulated. An obvious example involves making short quotations from a book (or other printed published work) for recognized purposes such as citing an authoritative source in a work of a kind that reasonably puts an onus on an author to provide evidence of veracity or credibility. Hence, says Lessig, with respect to Chapter 10 of *Free Culture*,

[y]ou are free to quote from this chapter, even in a review that is quite negative, without my permission, even though that quoting makes a copy. That copy would ordinarily give the copyright owner the exclusive right to say whether the copy is allowed or not, but the law denies the owner any exclusive right over such "fair uses" for public policy (and possibly First Amendment) reasons. (Lessig, 2004, p. 142)

To summarize the case of the book in relation to copyright law, we can follow Lessig in saying that in "real space"—the physical space of the book as an analogue artifact—the possible uses of a book fall into three categories: (1) unregulated uses, (2) regulated uses, and (3) regulated uses that are nonetheless deemed "fair" regardless of the copyright owner's views (Lessig, 2004, p. 143).

Figure 12.5. Change to Scope of Regulated Uses Within Digital Networks. Adapted from Lessig 2004, p. 143.

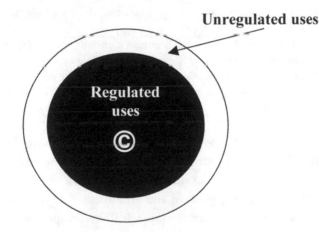

This example of the book from the physical, analogue world is overturned when the internet enters the equation. This is because the internet is "a distributed, digital network where every use of a copyrighted work produces a copy" (Lessig, 2004, p. 143). This "single, arbitrary feature of the design of a digital network" results in a dramatic change in the scope of unregulated uses (Category 1), whether we are talking about online books or any other kind of artifact that becomes a potential resource for writing (as remix). As the above figure illustrates, uses of kinds that hitherto were presumptively *unregulated* now presumptively become *regulated* (see Figure 12.5). There no longer remains "a set of presumptively unregulated uses that define a freedom associated with a copyrighted work." (Lessig, 2004, p. 143) Rather, every use now becomes subject to copyright, *because each use also makes a copy*. That is to say, the former

Category 1 uses (unregulated uses) automatically get sucked into Category 2 (regulated uses). As a consequence, "those who would defend the unregulated uses of copyrighted work must look exclusively to Category 3, fair uses, to bear the burden of this shift" (ibid., p. 143).

This has far-reaching consequences of diverse kinds. For example, before the advent of the internet, you could purchase a book and read it ten (or a hundred) times without the possibility of the copyright owner making a plausible *copyright*-related argument designed to control that use of his or her book. That is, copyright law "would have nothing to say about whether you read the book once, ten times, or every night before you went to bed. None of those instances of use—reading—could be regulated by copyright law because none of those uses produced a copy" (Lessig, 2004, pp. 143–144). But when this book takes the form of an e-book it effectively falls under a different set of rules.

> Now if the copyright owner says you may read the book only once or only once a month, then *copyright law* would aid the copyright owner in exercising this degree of control, because of the accidental feature of copyright law that triggers its application upon there being a copy. Now if you read the book ten times and the license says you may read it only five times, then whenever you read the book (or any portion of it) beyond the fifth time, you are making a copy of the book contrary to the copyright owner's wish. (ibid., p. 144)

Lessig recognizes that there are some people who think this makes perfect sense, although his immediate purpose is not to argue one side or the other on that point. Instead, he seeks simply to make as clear as possible the change in circumstances contingent upon the architecture of digital technology. Once the point is clear and grasped, some additional important points likewise become clear. Lessig spells out three crucial points (2004, pp. 144–145).

1. There is no evidence that policy makers intended to make Category 1 uses (unregulated uses) disappear when Congress allowed U.S. copyright policy to shift. Before the advent of the internet, unregulated uses of copyrighted works were an important part of free culture. It seems that Congress simply did not think through the collapse of the presumptively unregulated uses of copyrighted works and the consequences of this for free culture when it acted to change copyright policy.

2. The policy shift has particularly troubling consequences for *transformative* uses of creative content. Whereas the wrong involved in commercial piracy is easy enough for anyone to understand, the law would now seek to regulate any and all kinds of transformation individuals can make of a creative work through use of a machine. As a consequence, acts of copying, cutting, and

pasting become crimes. "Tinkering with a story and releasing it to others exposes the tinkerer to at least a requirement of justification. However troubling the expansion with respect to copying a particular work turns out to be, it is *extraordinarily* troubling with respect to transformative uses of creative work" (Lessig, 2004, p. 144; our italics).

3. The shift from unregulated (Category 1) uses to regulated (Category 2) uses puts enormous pressure on Category 3 ("fair use") uses as the sole basis for trying to maintain free cultural space. This is a burden that fair use has never previously had to bear, and that it is not well positioned to defend.

> If a copyright owner now tried to control how many times I could read a book online, the natural response would be to argue that this is a violation of my fair use rights. But there has never been any litigation about whether I have a fair use right to read, because before the Internet, reading did not trigger the application of copyright law and hence the need for a fair use defense. The right to read was effectively protected before because reading was not regulated. (ibid., p. 145)

A key problem that arises here concerns the fact that the point Lessig makes about fair use is often completely ignored, including by many free culture advocates. The prior question about the massive expansion in copyright regulation doesn't get addressed, because defenders of free culture have effectively "been cornered into arguing that our rights depend upon fair use." (ibid.) This is a very weak position to end up in. Whereas "a thin protection grounded in fair use makes sense when the vast majority of uses are *unregulated* . . . , when everything becomes presumptively regulated . . . the protections of fair use are not enough" (ibid.). This bespeaks the need to take up the prior question of the massive expansion in regulation and to not depend on "fair use" alone as the point from which to try and defend a space of free culture. For it cannot be equal to the task.

The Wider Ramifications for Digital Writing

The digital world, then, is fundamentally different from the analogue world from the perspective of the breach of copyright right law. To reiterate, the single most important and obvious fact about the digital world is that every single use of an artifact produces a copy. There is no way to access a work in a digital space and to even read it without producing a copy. It is the architecture of the digital environment that every single access or use produces a copy in some technical sense. Copies require permission. Therefore, if every single use creates a copy then it follows, in the copyright lawyer's mind, that *presumptively* every single use requires permission now. Hence, we go from that balanced picture of regulated and unregulated uses to the radically different picture *not* because any parliament or congress contemplated this change, but simply because of the interaction between the architecture of copyright law and the architecture of the digital network. We have, in effect, set up a world—or had a world set up for us—where permission is required for every single use (Lessig, 2004).

In a wide range of uses, of course, permission simply doesn't come. So, for example, Danger Mouse knew the Beatles never give permission to remix their work. The person who created *Tarnation* for $218 found it would cost over $400,000 to clear the rights to the music in the background of the video created from materials that he had shot when he was growing up. And when the Swedish Company, Read My Lips, that created the Bush and Blair love duet (see http://www.atmo.se/?pageID=4&articleID=389), contacted the copyright holders to ask permission to synchronise the images and the sound they received a very clear answer. When the lawyers denied Read My Lips permission to synchronize the sound and images that constitute the Bush-Blair love duet, they invoked the one idea that simply cannot be applied to the clip—regardless of anyone's particular views about Bush or Blair, or the Iraq War. The lawyers informed Read My Lips that they may not synchronize because, quote, "It is not funny." (Lessig, 2005a)

We have ended up in the situation where permission is required and permission does not come. The law has inverted the default of freedom that long existed in our culture. The consequence is that we have gone from a world where the regulation of culture was the exception to a world where the freedom of culture is now the exception. This change has happened because of changes in technology.

In fact, the situation is actually more complex and deep than has been stated so far. Examples like those of Danger Mouse and Johan Söderberg (of the Bush-Blair duet), along with legions of AMV creators, are examples of people engaging in freedom *despite* the rule of the law. What is much more profound, Lessig argues, is the change that technology is *about* to make happen if people in the content industry, along with Microsoft and certain others, prevail (2005a, 2005b). If technology companies become directed in the way the content industry wants them to be directed, the actual capacity to engage in that free use of culture may very soon be *removed* by the very same technology that presently enables it. This would be the consequence of having digital rights management technology layered into the network. Such technology would make it increasingly difficult and, for most people, impossible, to remix online digital resources. If this happens we will end up in a world where most people simply will not be able to engage in creative uses of culture through remix. Digital technology would thereby support the new default that the law has produced. Lessig argues that if and when this occurs we will have moved, in a short space of time, from a free culture to a permission culture—a culture in which the use of cultural resources critically and for extension will require the permission of the "culture owners" (Lessig, 2005b).

What does this mean? Lessig notes in the first instance that what it does *not* mean is that kids will not use technology to crack the technology that has been layered into content in the attempt to control free cultural activity. This is precisely what many of them *will* do. They will do it in order to be able to continue engaging in the type of cultural creativity known as digital remix. They will *always* engage in this type of creativity.

What it *does* mean, by contrast, is that we cannot formally *teach* them how to speak and write in this way. We won't be able to set up classes that engage in 21st century creative writing and learning in the kinds of ways we engaged in creative writing in the 20th century, because to do so would be to establish institutions and institutional practices that effectively support what the law currently calls "piracy." As educators we can't promote products and technology that engage in this form of community expression because, increasingly, to deploy products that enable people to engage in something called copyright violation is to make your own organization liable for a copyright violation.

At present what we are doing at the level of the law and corporate culture in response to these new technologies is punishing those who engage in digital remix as a form of creativity. We build tools to write, tools to understand the culture within which we all now live, and those tools *are rendered illegal*. Anyone reflecting on this fact would ask how it is that we produce this extraordinary conflict at the very moment that technology bestows the widest range of cultural opportunities to engage in expressive activity. Why would the law come in and take the freedom to engage in that expressive activity away?

The obvious answer is that any technology that enables creative forms of expression or of remix also enables what's come to be known as "piracy." As Lessig notes in different contexts (Lessig 2004, 2005b), the President of the Motion Picture Association of America, Jack Valenti, has referred to his own private "terrorist war." It seems that we have children who are the terrorists here (Lessig, 2005b). This situation reflects an inability to think subtly about complex issues when these complex issues are captured by a political system deeply controlled by the most powerful sectional interest lobbies within popular culture (Lessig, 2005b). This kind of extremism is predictable. Furthermore, it has produced an explosion of "weapons" for dealing with this piracy, in the form of new laws and technologies (e.g., digital rights management, or DRM, code added to digital artifacts like music and movies) to protect intellectual property.

The point that most needs to be made and recognized here is that these new laws and these new technologies designed to protect the architecture of revenue from the 20th century will have the consequence of destroying the potential of this technology for producing something radically new in the ways ordinary culture gets made and shared. The existing law of copyright fundamentally conflicts with these new technologies and forces a choice on us of either reforming the law or reforming the technology. In the United States, especially, and in the European Union and elsewhere, the response to this conflict to date is: We will reform the technology.

Lessig argues that we need as a culture to seek a more mature response. He emphasizes that this response is not against copyright. Indeed, Lessig identifies himself as "fundamentally a believer in the system of copyright." (2005b, no page) He regards copyright as essential in the digital age—he believes it is as essential as it was in the analogue age. At the same time he maintains that what the world needs is to recognize what lawyers have understood for the last several centuries. This is that *there is no single system of copyright*. Rather, there is a system of copyright that evolves as technology evolves, and it changes dramatically over a very short period of time in light of a new technology. The current system of copyright is not bad *because it is a system of copyright*. It is bad because it is simply out of date with the technology. We need merely to update it to cohere with the digital technology that now pervades our life. We need to make it make sense of this digital technology. We need to find a way to get those who have the power to do so to begin to reform the law and to make what was always free—namely remix—free again in the 21st century.

And this presupposes reform.

Steps Toward Reform

The difficult question here is: How do we get to the place where reform is possible, where reform can reflect these changes in as mature a way as possible?

Lessig suggests that four steps are essential for getting us there.

1. Most importantly, those seeking reform in the interest of the right to write and create freely yet responsibly in ways that honor the need to reward original creators fairly need to find ways to connect to those who support the current legal arrangements. That means we must be willing more frequently and openly to call *piracy* "piracy." If the war that is being waged right now really were about the right to access music for free, it would be a war that is not worth waging. Getting access to owners' copyrighted material against their wishes is not what free culture is about. We need to be willing to say that something wrong is being done by those who use this technology to violate other peoples' rights. We are not defending that use. We think it is wrong. Rather, what we are defending is the freedom and opportunity that could be built by a world where people are free to use these technologies in ways that historically we have always used words: to create culture (Lessig, 2005a, 2005b).

2. We need to educate copyright lawmakers about how extraordinarily powerful this technology is for all the things free societies espouse and value. We can educate them by showing precisely what our young people do with this technology. This is not about just building huge archives of every song ever recorded. Rather, it is about increasingly finding ways to use these bits of our culture to speak differently, more powerfully, more effectively (Lessig, 2005a, 2005b).

3. We need to find ways to encourage change in the laws that are so radically out of date with these new technologies. We need to educate lawmakers about how to change copyright law, not how to abolish it. This is not a call for the end of intellectual property. It is a call for the practice that has been the history, at least of copyright regulation, to update it in ways that bring it into a reasonable alignment with the technology (Lessig, 2005a, 2005b). This calls for voluntary action by people around the world who can begin to symbolize and create a different environment for creativity. This is the objective of the non-profit organization that Lessig founded and runs called Creative Commons (see Creativecommons.org).

Creative Commons aims to create technologies that produce a simple way for authors and artists to mark their content with the freedoms they intend their content to carry. This informs would-be users/remixers of this work of the conditions under which they are permitted by the original creator to use this work. When creators go to the Creative Commons website they are given a choice of licenses to apply to their work. They can select among licenses that allow creators to permit or constrain commercial uses of their work, to permit or constrain modifications of their work and the conditions under which those who modify a creator's work are entitled to release any work that builds on that creator's work. So, for example, if a creator allows modifications of their work they may—if they wish—select a license requiring that other creators release any work that builds upon their own work in similarly free terms.

That selection process among the different available options produces a license—"some rights reserved"—governing subsequent cultural uses and remixes of a particular creator's work. For example, the copyright license governing the digital version of Lessig's book, *Free Culture* (2004), is a "Creative Commons Attribution-Noncommercial 3.0 United States License" (Lessig. org). What this means is that

"You are free:
  • to Share—to copy, distribute, display, and perform the work
  • to Remix—to make derivative works

Under the following conditions:
  • Attribution. You must attribute this work to Lawrence Lessig (with link).
  • Noncommercial. You may not use this work for commercial purposes.
  • Permissions beyond the scope of this public license are available at www.randomhouse.com.
  • For any reuse or distribution, you must make clear to others the license terms of this work. The best way to do this is with a link to this web page.
  • Any of the above conditions can be waived if you get permission from the copyright holder.
  • Apart from the remix rights granted under this license, nothing in this license impairs or restricts the author's moral rights." (Source: http://creativecommons.org/licenses/by-nc/3.0/us/)

Creative Commons licenses come in three separate layers. The first layer is a humanly readable common deed. This is the expression of a freedom associated with that content in terms that anyone should be able to understand. The second layer is a lawyer-readable license. This is intended to make enforceable the freedoms associated with that content. The third layer—and the layer Lessig regards as most important—is a machine-readable legal expression of the freedoms associated with that kind of content. This layer means that search engines could begin to gather content on the basis of the freedoms involved.

So, for example, potential users of cultural resources can direct a search engine to "Show me all the pictures of the Empire State Building that are available for noncommercial use" and the search engine will gather the relevant works on the basis of freedoms.

Creative Commons licenses have spread at an increasingly rapid rate. In the first year around one million licenses were issued. After 18 months the number had reached 1.8 million, and by the end of the second year it was over 4 million. By August 2005, according to Lessig himself, the number was 53 million. The objective of this initiative is to facilitate the very obvious creativity the internet enables, and to date many millions of cultural creators have responded to this aim.

The key point to recognize here is that this kind of creativity is encouraged legally, consistent with copyright law, without any lawyers standing between the creators: that is, between the original creators and those who would subsequently use and transform these creators' works in their own acts of cultural creativity. No one speaks directly to each other. No one needs to be negotiating anything. The freedom to engage in creativity is built into the licensing architecture. By this means we confront a world where the copyright machine has layered itself on to the internet in a way that forces permission everywhere and, voluntarily, artists will add a layer of freedom on top of this system to facilitate something like collaboration that can occur legally.

Lessig identifies Creative Commons as a step toward creating this kind of environment for creativity. But he does not see it as the solution. The solution requires much more than this kind of voluntary project, but it points to a certain kind of urgency that Lessig reflects upon. This is an urgency that is particularly alien to those—like Lessig himself—who try to teach people about the virtue and importance of the rule of law. The fact is that the new tools for creativity are here to stay. No one is going to abolish them. Supporters of the current copyright law regime may succeed in forcing this creativity to live underground and, indeed, a war on technologies will produce an underground culture. This will be one pole of a culture that is deeply alienating, a culture that increasingly pushes a kind of extremism in this debate.

Such extremism creates doubt. Lessig argues that both extremes in front of us today—the underground culture of violation and the establishment culture of copyright extremism—are wrong (Lessig, 2005b). Both extremes are harmful because they produce a generation that thinks the law is an ass and that the law is, therefore, to be ignored. It is terrifying to see the kind of attitude of righteous law violation that spreads in this kind of context, where the law expresses itself in such an extreme form that reasonable, sensible, intelligent, innovative kids think that the only proper response is to violate the law.

In a democracy this makes no sense. In a democracy we should be able to reform the way the law works so most people think that this law makes sense, and at present this law does *not* make sense, because what we need now is a movement that begins to recognize the insanity in this extremism and restore some sense of soundness to the debate. We need it now if we are going to produce a world where writing is allowed and encouraged in all the ways in which our culture has traditionally encouraged writing. This, of course, will be writing in a sense that the 21st century, with its sophisticated and rapidly evolving digital technologies now makes possible.

4. Finally, then, we have got to find ways to deliver our message effectively to the copyright extremists. This calls for political mobilization. We have got to establish clear messages about what freedom means, and we have to oppose those who disagree with these messages and to defend this freedom.

Conclusion

Overly regulated copyright laws in the digital domain risk creating a "read-only" world for almost everyone. We have to defend the freedom to remix media under copyright laws that respond to respectful uses of other people's ideas, materials, and creativity where "respectful use" is defined by the creators themselves. We have to change the law, because otherwise they—copyright lawmakers and copyright enforcers—will destroy our new and emerging productive technologies, and our right to create freely, but responsibly and respectfully, with them.

The significance of the challenge for education is acute, on a number of dimensions, and the need to take it up is urgent.

For example, it is important to recognize the extent to which politicians, education policy makers, education administrators, teacher educators, and teachers can currently take shelter behind fears and confusions associated with copyright and permissions to "justify" maintaining educational business as usual. This is an almost inevitable consequence of the traditional and entirely outmoded conception of education as content delivery. Where education is seen as being about transferring content we need to be very careful that we do not transgress against copyright and intellectual property ownership. Where particular content is regarded as so important that its transmission is believed to constitute the true purpose of education, individuals and corporations will be keen to own it so that they can derive economic and social rewards from its (compulsory) educational use. Whence, the textbook industry and the stranglehold it exerts over school education.

By contrast, a new generation of learning theorists and other educationists is doing their best to convince anyone open to the idea that education is best understood in terms of pursuing capacity for expert performance in social practices and discourses that enable us to live as well as possible in the world as social, economic/professional, civic, aesthetic, ethical, spiritual/emotional, etc. beings. This entails interacting with cultural tools and artifacts, with norms and rule systems, and with other people (experts, peers, novices) within situations and contexts that approximate as closely and appropriately as possible to those in which learners will find themselves throughout their subsequent life trajectories. This is not about content transfer. It is about acting in and on the world using cultural artifacts. It is, in other words, very much like Lessig's idea of learning to write by manipulating existing cultural material. This is not about plagiarism, since the point is not to create a product per se. It is *only* when content transfer and the demonstration of that transfer are mistaken for ends of education that plagiarism becomes a problem. Any educational task that can seemingly be met by an act of plagiarism is not an adequate educational task. On the other hand, the kind of cultural remixing that takes existing products, puts them together as a way of seeing how they work and what kinds of things result from putting them together in different ways, becomes part of a progressive practice of learning how to do what experts do, by leaning on them through the medium of their cultural creations. Slowly but surely, as understanding of areas of inquiry and social practices of diverse kinds are "unveiled" and understood in terms of systems of norms and rules, and standards (and the modifiable, evolving and "improvable" nature of these) become enhanced through purposeful situated engagements, learners lean less comprehensively on intact cultural creations and increasingly generate their "own"—albeit as remixes to a greater or lesser extent of extant cultural materials they encounter in their environments (e.g., Bransford, Brown & Cocking, 1999; Brown, Collins & Duguid, 1989; Gee, 2003, 2004, 2007; Squire, 2006, in press).

Similarly, the status quo with respect to copyright, permissions, and intellectual property abets those whose comfort zones demand that literacy be defined as closely as possible and for as long as possible in terms of alphabetic print and the book as the text paradigm. Bookspace—after all these centuries—is comfortable space. It marks out a comfort zone. We know where we stand with books, so to speak. We know how to operate with them and we don't have to live in fear of risks we might unwittingly entertain by going into unfamiliar territory. This also, of course, works well for textbook publishers and for curriculum developers who can stay close to familiar territory without having to learn new tricks.

The moment we begin to address free culture in terms of the right to write as remixers, and in terms of opportunities to make *experience* of cultural creativity through the capacity to mix and manipulate cultural materials from everyday life—whether these be finite artifacts, or whether they be symbols, elements of theories, or designs—is the moment we begin to challenge seriously the prevalent conception of education as content transmission. That will be the moment we begin seriously to de-commodify education and reconstitute learning as an expression of free cultural production in the interests of becoming expert performers within those domains of everyday life that education should properly be concerned about (cf. Illich, 1970; Gee, 2004, 2007; Lankshear and Knobel, 2006, Chapter 8).

We personally believe that this moment cannot come too soon.

References

Bransford, J., Brown, A., & Cocking, R. (1999). *How people learn*. Washington, DC: National Research Council.

Brown, J. S., Collins, A., & Duguid, P. (1989). Situated cognition and the culture of learning. *Educational Researcher* 18(1), pp. 32–42.

Gee, J. (2003). *What video games have to teach us about learning and literacy*. New York: Palgrave.

———. (2004). *Situated language and learning: A critique of traditional schooling*. London: Routledge.

———. (2007). *Good video games + good learning: Collected essays on video games, learning and literacy*. New York: Peter Lang.

Illich, I. (1970). *Deschooling society*. Harmondswoth, UK: Penguin.

Knobel, M. and Lankshear, C. (2008). Remix: The art and craft of endless hybridization. *Journal of Adolescent and Adult Literacy*. 52 (2).

Lankshear, C. and Knobel, M. (2003). *New Literacies: Changing Knowledge and Classroom Learning*. 1st Ed. Buckingham, UK: Open University Press.

———. (2006). *New literacies: Everyday practices and classroom learning*. Second edition. Maidenhead, UK: Open University Press.

Lessig, L. (2004). *Free culture: How big media uses technology and the law to lock down culture and control creativity*. New York: The Penguin Press. Available: http://www.free-culture.cc

———. (2005a). Keynote address presented at the ETech Emerging Technology Conference, San Diego, 17 March 2005.

———. (2005b). Re:MixMe. Keynote address presented at the ITU "Creative Dialogues" Conference, University of Oslo, 21 October 2005.

Squire, K. (2006). From content to context: Video games as designed experiences. *Educational Researcher* 35(8), pp 19–29.

———. (in press). From information to experience: Place-based augmented reality games as a model for learning in a globally networked society. Forthcoming in *Teacher's College Record*.

Stallman, R. (2002). *Free software, free society: Selected essays of Richard M. Stallman*. Boston: GNU Press/Free Software Foundation.

Contributors

David Bawden: David is Professor of Information Science at City University London. He is also editor of *Journal of Documentation*, the leading European academic journal of the information sciences. His research interests include the foundations of the information sciences, information history, and information behavior, as well as digital literacies.

David Buckingham: David is Professor of Education at the Institute of Education, London University. He is the found and director of the Centre for the Study of Children, Youth and Media. He is the author, co-author or editor of 22 books and well over 180 articles and book chapters. His work has been translated into 15 languages.

Julia Davies: Julia is based at The University of Sheffield in the UK, where she directs the online MA in New Literacies and MA in Educational Research programmes. Her research focuses on the relationship between social learning and online digital practices, online identities and the nature of digital texts.

Forthcoming work includes a book with Guy Merchant (2008) titled *Web 2.0 for Schools: Learning and Social Participation*.

Lilia Efimova: Lilia is a researcher at Telematica Instituut, Netherlands. Her interests include research on personal knowledge management, passion at work and social media, as well as combining research with hands-on experience in facilitating learning and change. Lilia writes at blog.mathemagenic.com.

Ola Erstad: Ola is Associate Professor and Head of Research at the Department for Educational Research, University of Oslo, Norway. He works within the fields of both media and educational research. He has published on issues of technology and education, especially on "media literacy" and "digital competence." Among recent publications is a chapter written with James Wertsch on "Tales of Mediation: Narrative and Digital Media as Cultural Tools" in a forthcoming book edited by K. Lundby (Peter Lang).

Maggie Fieldhouse: Maggie is a Lecturer at the School of Library, Archive and Information Studies (SLAIS) at University College, London, where she teaches in the Library and Information Studies Programme. Her research interests include pedagogical aspects of information literacy and the relationship between digital literacy and information-seeking behavior.

Jonathan Grudin: Jonathan is Principal Researcher at Microsoft and Affiliate Professor at the University of Washington Information School in the USA. His research interests include the adoption of technologies, especially in enterprise settings. He was editor-in-chief of *ACM Transactions on Computer-Human Interaction* and is *the ACM Computing Surveys* associate editor for *Human Computer Interaction*.

Genevieve Johnson: Genevieve is Professor of Psychology at Grant MacEwan College, Edmonton, Canada. Her program of research includes theoretical, empirical, and practical understandings of internet technologies and human learning and cognition. Increasingly complex cultural tools (e.g., the internet) require, reflect, and facilitate increasingly complex cognitive processes (and vice versa). For a complete list of Genevieve's publications, visit http://members.shaw.ca/gen.johnson

Michele Knobel: Michele is a Professor of Education at Montclair State University (U.S.), where she co-ordinates the graduate literacy programs. Her research focuses on the relationship between new literacies, social practices and digital technologies. She blogs, with Colin Lankshear, at: http://everdayliteracies.blogspot.com. Recent books include the *Handbook for Teacher Research* and *A New Literacies Sampler* (both with Colin Lankshear).

Colin Lankshear: Colin is Professor of Literacy and New Technologies at James Cook University (Australia) and Visiting Scholar at McGill University (Canada). His research interests include philosophical and empirical investigation at interfaces between new technologies, literacy and everyday social practices. He is co-author (with Michele Knobel) of *New Literacies: Everyday Practices and Classroom Learning* and joint editor of the *Handbook of Research on New Literacies*.

Allan Martin: Allan has degrees in Sociology, Folklore and Education. He taught in secondary schools for ten years, and then in teacher education, returning to Glasgow (his birthplace) in 1995 to develop the University's IT Literacy Programme, which he led until his retirement in 2007. He continues to write and research on digital literacy and education. He edited, with Dan Madigan, *Literacies for Learning in the Digital Age*. He is currently a member of the European Commission's Expert Group on Digital Literacy.

David Nicholas: David is Professor and the Director of the School of Library, Archive and Information Studies (SLAIS) at UCL. He is also the Director of the UCL Centre for Publishing and Director of the CIBER research group. His research interests largely concern the virtual scholar and he is currently engaged in investigations of the use and impact of e-journals and e-books.

Leena Rantala: Leena is a doctoral student and researcher in the Department of Education, University of Tampere, Finland. Her. Ph.D. study focuses on digital literacy practices in school culture. Her research interests include media and civic education, new literacies and methodological questions related to critical ethnology.

Mortan Søby: Morten is Head of Section at University of Oslo, Norwegian Network for IT-Research and Competence in Education and has been the executive editor of the *Nordic Journal of Digital Literacy* since 2006. His research interests include digital literacy, web 2.0 and ICT policy.

Juha Suoranta: Juha is Professor of Adult Education (Finland). His research interests include theoretical and methodological questions in critical pedagogy and media education. He is co-author (with Tere Vadén) of *Wikiworld: Political Economy of Digital Literacy and the Promise of Participatory Media.*

Names Index

Subject Index

A

assessment 41–2, 67, 131, 136, 145, 147
  of functional internet literacy 41–2
authoring in remix 187–90

B

baby boomers 56, 67
basic competence 109, 131, 132–4
  and *Bildung* 134
  categories of 133
  as plural 1–2
  sociocultural view of 4–7
blogging (see also weblogging) 5–6, 13, 14,
  80, 85, 125–6, 163, ch 9
Bloom's taxonomy of cognitive skills 37–8,
  42

application to functional internet literacy
  37–40

C

cognitive skills 37–40
  Bloom's taxonomy of 37–8, 42
  and functional internet literacy 37–40
computer literacy 4, 17, 21, 22–5, 29, 33, 41,
  76, 95, 107, 156–7, 167
computer supported collaborative learning
  128–9
constructions of reality through language
  243–5
copyright 55, 58–9, 65, 186, 260–61, 280–81,
  289–90, 291–304
  challenge for education 303–4
  and democracy 302

Colin Lankshear, Michele Knobel,
& Michael Peters
*General Editor*s

New literacies and new knowledges are being invented "in
the streets" as people from all walks of life wrestle with
new technologies, shifting values, changing institutions,
and new structures of personality and temperament emerging
in a global informational age. These new literacies and
ways of knowing remain absent from classrooms. Many education administrators, teachers, teacher educators, and academics seem largely unaware of them. Others actively
oppose them. Yet, they increasingly shape the engagements
and worlds of young people in societies like our own. The
New Literacies and Digital Epistemologies series will explore this terrain with a view to informing educational
theory and practice in constructively critical ways.

    For further information about the series and submitting
manuscripts, please contact:

                Michele Knobel & Colin Lankshear
                Montclair State University
                Dept. of Education and Human Services
                3173 University Hall
                Montclair, NJ 07043
                michele@coatepec.net

    To order other books in this series, please contact our
Customer Service Department at:

                (800) 770-LANG (within the U.S.)
                (212) 647-7706 (outside the U.S.)
                (212) 647-7707 FAX

Or browse online by series at:
                www.peterlang.com